Organic Substances in Soil and Water:
Natural Constituents and Their Influences on Contaminant
Behaviour

Organic Substances in Soil and Water: Natural Constituents and Their Influences on Contaminant Behaviour

Edited by

A. J. Beck and K. C. Jones,
Environmental Science Division, Lancaster University, UK

M. H. B. Hayes
School of Chemistry, University of Birmingham, UK

U. Mingelgrin
Department of Physical and Environmental Chemistry, Institute of Soils and Water, Volcani Center, Bet Dagan, Israel

ROYAL
SOCIETY OF
CHEMISTRY

Based on the Keynote Papers presented at the International Conference on Organic Substances in Soil and Water held at Lancaster University, UK on 14–17 September, 1992

Special Publication No. 135

ISBN 0-85186-635-2

A catalogue record of this book is available from the British Library.

Published by The Royal Society of Chemistry.
Thomas Graham House, Science Park, Cambridge CB4 4WF
Typeset by Keytec Typesetting Ltd., Bridport, Dorset
Printed by Hartnolls Ltd., Bodmin

SD 2/2/94

Preface

Soils are widely known to comprise both a source and a sink for a wide range of compounds in the environment. These may be of natural or anthropogenic origin and may be of an inorganic or organic nature. A number of these compounds, such as pesticides, polychlorinated biphenyls, and some polynuclear aromatic hydrocarbons, in addition to toxic heavy metals, have been designated priority pollutants by the United States Environmental Protection Agency and the European Commission because of their known, or suspected toxicity. Therefore, it is vital that their behaviour in the environment be understood if we are to minimize the risk they pose to both our biotic environment and ourselves.

Soils contain a wide range of natural organic substances derived from the decomposition of animal and vegetable matter, and from the application of waste materials such as farmyard manure and sewage sludge. These substances occur as solids, in various stages of humification, in organic rich surface horizons, or as cutans on clay minerals further down soil profiles. As water moves through the soil this organic material may be subjected to solubilization resulting in soil solutions of complex chemical compositions. It has been known for some time that the behaviour of many compounds is strongly influenced by the solid- or stationary-phase organic matter in soils. More recently the mobile- or dissolved-phase organic matter has also been implicated as being of importance in governing the behaviour of many compounds and particularly those that are extremely hydrophobic. In this book we assess the 'state of the art' in this field of research by drawing together contributions from leading authorities on the nature of solid- and dissolved-phase organic substances, their interactions with organic and inorganic chemicals, and the implications of these interactions for agriculture and environmental pollution.

It is widely regarded that the most important attribute of soil is as a medium for food production. In the first chapter Johnston examines the importance of soil organic matter in this context. He then explores the consequences for food production of manipulating soil organic matter levels to reduce mobility of compounds used in crop protection that are perceived to be pollutants in water bodies. The need to identify the source and nature of the dissolved organic matter involved in transport processes is stressed as a prerequisite to any such activity. Chapters 2 and 3 provide some answers to these questions. In the second chapter Malcolm reviews the concentration and composition of dissolved organic matter not only in soil interstitial water but also in streams and groundwaters and stresses the differences between them. The problems of equating fulvic and humic acid components with dissolved organic carbon (DOC) are highlighted and the scientific community is incited to pursue research on the composition and implications of hydrophilic and hydrophobic neutral fractions of DOC. In Chapter 3 Hayes and colleagues summarize contemporary techniques which are being used to isolate and characterize the structures of organic macromolecules in soils. Non-destructive techniques, including cross-polarization magic angle spinning (CPMAS) ^{13}C NMR, Fourier transform infrared (FTIR), Mössbauer, and electron spin resonance (ESR) spectrocopy, are discussed in addition to chemical and pyrolysis degradation techniques and physico-chemical procedures.

In Chapters 4–7 the emphasis of discussion is placed on the interaction between organic substances and other natural or anthropogenic substances, many of which are perceived to be pollutants. In Chapter 4 Senesi provides a detailed discussion of the multitude of chemical and physical interactions that are possible between humic substances and a diverse range of xenobiotic compounds. These include chemical and physical binding, hydrophobic adsorption and partitioning, enzyme mediated binding, solubilization effects, and photosensitization. The effect of environmental factors including pH, ionic strength, temperature, moisture, and clay on these interactions are also considered. Investigations of organic matter interactions with pollutants have become somewhat polarized, with a large body of the research community studying the solid phase and a smaller group focusing on the mobile dissolved phase. Such an approach may be counterproductive where the behaviour of interest is on the larger scale. In Chapter 5 Mingelgrin and Gerstl demonstrate that a unified approach to the interaction of pollutants with organic matter is possible. However, they stress that our understanding of kinetic factors is likely to limit our success in quantitatively predicting equilibrium sorption, and more so transport, availability, and susceptibility to biotic degradation where the dynamics of sorption and desorption are of greater importance than the equilibrium state. The importance of kinetics has been manifest in many recent publications by Brusseau and Rao, Ball and associates, and by Pignatello and co-workers. In Chapter 6 Dr Pignatello reviews some recent advances in our understanding of sorption/desorption kinetics for organic chemicals in soils and sediments and highlights unresolved problems requiring further study. He emphasizes the timescale of reactions and discusses the implications of these for transport, bioavailability, and remediation. He also discusses the methodological difficulties involved in investigating kinetics. One novel approach to this problem is presented in Chapter 7 by Gamble and associates who outline a research strategy they have developed to produce stoichiometrically exact interpretations of the kinetics for pesticide and metal interactions with soils.

All of the preceding issues are integrated and applied to the field scale in Chapter 8 in which McCarthy discusses the subsurface transport of dissolved organic matter and its implications to contaminant mobility. He echoes Johnston's appeal to identify the organic fractions of interest by stressing the need to recognize the heterogeneity of dissolved organic matter with regard to contaminant behaviour, thus encouraging further research previously suggested by Malcolm. Chapter 9 is devoted to modelling. In it Professor Bengtsson reviews the success of various mathematical models to simulate sorption/desorption kinetics and enzyme-reaction kinetics. The need to better understand the processes being modelled and to establish experiments which can be used for validation and verification are emphasized. In the final chapter we consider the advances that have been made in our understanding over recent years and highlight unresolved problems that will require further research over the coming decade.

A. J. Beck (Lancaster)
K. C. Jones (Lancaster)
May 1993

Contents

Organic Matter in Soils and Waters: Significance, Distribution, Composition, Structure

1 Significance of Organic Matter in Agricultural Soils
   A. E. Johnston   3

2 Concentration and Composition of Dissolved Organic Carbon
   in Soils, Streams, and Groundwaters
   R. L. Malcolm   19

3 Isolation, Fractionation, Functionalities, and Concepts of
   Structures of Soil Organic Macromolecules
   C. E. Clapp, M. H. B. Hayes, and R. S. Swift   31

Interactions between Contaminants and Naturally Occurring Organic Substances

4 Nature of Interactions between Organic Chemicals and Dissolved
   Humic Substances and the Influence of Environmental Factors
   N. Senesi   73

5 A Unified Approach to the Interaction of Small Molecules with
   Macrospecies
   U. Mingelgrin and Z. Gerstl   102

6 Recent Advances in Sorption Kinetics   128
   J. J. Pignatello

7 Interactions of Organic and Inorganic Contaminants with Soils:
   Unifying Concepts
   D. S. Gamble, C. H. Langford, and G. R. B. Webster   141

8 Sub-surface Transport of Natural Organic Matter: Implications for
   Contaminant Mobility
   J. F. McCarthy   153

9 Modelling Solute–Sorbent Interactions of Saturated Flow in
   Heterogeneous Soils   171
   G. Bengtsson

10 Natural Organic Substances and Contaminant Behaviour:
    Progress, Conflicts, and Uncertainty   184
    A. J. Beck and K. C. Jones

Subject Index   195

Organic Matter in Soils and Waters: Significance, Distribution, Composition, Structure

1

Significance of Organic Matter in Agricultural Soils

By A.E. Johnston

AFRC IACR ROTHAMSTED EXPERIMENTAL STATION, HARPENDEN, HERTFORDSHIRE, AL5 2JQ, UK

1 Introduction

In his report[1] entitled 'Organic Chemistry in its Application to Agriculture and Physiology' presented in 1840 to the British Association for the Advancement of Science, Liebig discussed the view then held by vegetable physiologists that humus, produced by the decomposition of vegetable matter, was the principle nutriment of plants being directly extracted by plant roots from the soil. Although Liebig successfully demolished this hypothesis, he nevertheless proposed an important role for humus. 'Humus does not nourish plants by being taken up and assimilated in its unaltered state, but by presenting a slow and lasting source of carbonic acid which is absorbed by the roots and is the principal nutriment of young plants at a time when, being destitute of leaves, they are unable to extract food from the atmosphere'. Later he noted, 'Its (humus) quantity heightens the fertility of a soil ...'. Liebig was aware that the breakdown of organic matter produced carbon dioxide and he later proposed that carbonic acid was responsible for releasing elements like potassium and magnesium from soil minerals.

The concept that the fertility of a soil depended on its humus content was soon disproved by the results of field experiments on crop nutrition started by Lawes and Gilbert in 1843 at Rothamsted. They showed that nutrients like nitrogen, phosphorus, and potassium were taken up by roots from soil, and that adding organic matter to soil to produce carbon dioxide had no benefit. As early as 1845, it had been demonstrated that it was more important to have within soil a supply of 'available and assimilable nitrogen' to increase yields of winter wheat and turnips.[2] Subsequently Lawes and Gilbert often commented that although their experimental soils often contained 4000 kg ha^{-1} total nitrogen (in organic matter) it was in a form which did not benefit crop growth because when up to 100 kg ha^{-1}

nitrogen was given as a water soluble ammonium or nitrate salt there was a very large increase in yield. However, 130 years later, Russell noted that there has always been a lingering concern about the relationship between the fertility of a soil and the quantity of humus, that complex of organic compounds found mainly in the surface soil. The paper from which this comment was taken[3] was probably, in part, a considered response to the Strutt Report on 'Modern Farming and the Soil'.[4] This report was commissioned following two very poor harvests in consecutive years, 1968 and 1969. The enquiry was an official response to vigorously expressed, but poorly identified, concerns that modern farming, principally the intensification of arable cropping, mainly in the eastern counties, was having adverse effects on soil fertility. Soil humus content was one factor considered and the report suggested that humus levels should not be allowed to fall below about 3%.

In his paper[3] Russell went on to point out that the problem facing agricultural researchers was to define the role of humus in soil fertility, the time scales and husbandry practices which affect humus content, and the separation and quantification of the various factors which could contribute to any overall humus effect. However, as often happens in many branches of science, it is essential to define the problem to encourage research. This has been so with the agronomic effects of humus; until there were measurable effects on yield few researchers were interested in quantifying the various components which contribute to the overall effect. In temperate climates soil organic matter levels change only slowly and it may take many years to get soils with appreciably different levels of humus in an experiment where its effects can be estimated with accuracy. In such experiments it is also essential to ensure that no other factor like soil acidity or phosphorus (P) and potassium (K) status will limit growth. In tropical climates soil organic matter levels change much more rapidly but humus effects may be masked by chronic water shortage and organic additions to soil may be better left as surface mulches to minimize the risk of soil erosion.

This paper considers some aspects of the significance of humus in the productivity of temperate soils using data from the Rothamsted and Woburn Experimental Stations. Rothamsted dates its foundation as 1843, most of the soils are silty clay loams, 25–30% clay, and average annual rainfall is about 700 mm. Woburn was started in 1876 by the Royal Agricultural Society of England; it has been managed by Rothamsted since 1926. The soils at Woburn are mainly sandy loams, about 10% clay, and average annual rainfall is 650 mm. On both farms there are experiments with contrasted treatments which have affected the humus content of the soil. Equally important is the existence of a unique archive of crop and soil samples which have been taken throughout the history of these experiments. This archive allows estimates of many organic and inorganic constituents to be made now so that changes in concentration over time can be studied.

2 Effect of Soil Organic Matter on Crop Yield

All the early experiments at Rothamsted compared yields given by plots treated with either inorganic fertilizers or farmyard manure (FYM). Yields of winter wheat, spring barley, and mangolds were as large on plots given small quantities of readily available plant nutrients in fertilizers as on plots with much larger amounts of FYM (Table 1). The similarity of the yields of these three crops, together with those of sugar beet, on the two differently treated soils continued into the 1960s and 1970s, even though there was by this time about 2.5 times as much humus in FYM- as in fertilizer-treated soils. The difference in soil organic matter had arisen because of the long continued annual applications of FYM.

On the sandy loam at Woburn beneficial effects from extra soil organic matter were suspected in the Market Garden experiment in the early 1960s.[5] This led to a number of experiments testing organic matter, see for example Mattingly *et al*.[6] Results from one experiment in the 1970s showed that potatoes and spring barley benefited from extra humus but not winter wheat and winter barley (Table 2). This difference suggested that whilst deep rooted, autumn sown crops could make better use of subsoil water, spring sown, shallow rooted crops might be more responsive to differences in soil organic matter. The benefit could be through better soil structure and/or a small increase in the soil's water holding capacity. Although the latter might not amount to more than a few days water use by the crop this could be sufficient to delay the onset of severe water stress before the next rainfall. Another very important result in Table 2 is that for both potatoes and spring barley extra fertilizer nitrogen did not substitute for less soil organic matter. For both crops the final increment of nitrogen tested did not give an increase in yield on a low organic matter

Table 1 *Yields,* $t ha^{-1}$, *of winter wheat and spring barley grain, at 85% dry matter, and roots of mangolds and sugar beet at Rothamsted*

Experiment	Crop	Period	Yield with FYM	Yield with NPK fertilizers*
Broadbalk	Winter wheat	1852–61	2.41	2.52
		1902–11	2.62	2.76
		1970–75	5.80	5.47
Hoosfield	Spring barley	1856–61	2.85	2.91
		1902–11	2.96	2.52
		1964–67	4.60	3.36
		1964–67	5.00	5.00†
Barnfield	Mangolds	1876–94	42.2	46.0
	Mangolds	1941–59	22.3	36.2
	Sugar beet	1946–59	15.6	20.1

*FYM, $35 t ha^{-1}$; N to winter wheat, $144 kg ha^{-1}$; to spring barley, $48 kg ha^{-1}$; (except †, $96 kg ha^{-1}$); mangolds and sugar beet, $96 kg ha^{-1}$.

Table 2 *Yields of potatoes, winter wheat, and winter and spring barley at Woburn 1973–80*

% C in soil	Fertilizer N applied*				Average April–July rainfall and deviation from long-term average†
	0	1	2	3	
Potatoes, tubers, t ha⁻¹, *1973 and 1975*					
0.76	25.7	35.6	41.7	43.2	
2.03	27.1	40.6	50.7	59.0	266(−14)
Spring barley, grain, t ha⁻¹, *1978*					
0.76	2.19	5.00	6.73	7.05	
1.95	2.58	5.12	6.85	7.81	222(+20)
Winter wheat, grain, t ha⁻¹, *1979*					
0.76	3.54	7.32	8.05	7.82	
1.95	4.81	7.21	8.09	8.08	247(+45)
Winter barley, grain, t ha⁻¹, *1980*					
0.76	3.05	6.01	7.32	7.83	
1.95	3.57	5.92	7.00	7.98	258(+56)

* N0, N1, N2, N3: 0, 100, 200, 300 kg N ha⁻¹ for potatoes; 0, 50, 100, 150 kg N ha⁻¹ for cereals.
† April–August rainfall for potatoes.

soil. Much of this nitrogen could have been left in soil as nitrate in autumn and hence prone to loss by leaching. Further examples of the benefits of extra humus on the sandy loam at Woburn have been given elsewhere.[7]

On the silty clay loam at Rothamsted it took somewhat longer to demonstrate beneficial effects of extra humus. In the early 1970s yields from the two ley-arable experiments showed no beneficial effect from extra organic matter accumulated during a three-year period of grass, grass-clover, or lucerne leys. Very similar yields of winter wheat, potatoes, and spring barley were obtained after the leys and in an all-arable rotation, provided the correct amount of fertilizer nitrogen was used. But, the optimum nitrogen application was always less after a ley than in an all-arable rotation, because of the nitrogen released as the organic matter from the ley decomposed. Yields of potatoes were generally larger, however, in the experiment on soils with 3.6% organic matter than in the experiment on soil with 2.7% organic matter. This effect of extra humus on yields of potatoes was shown more convincingly, together with a similar benefit for sugar beet but not cereals, in another experiment at Rothamsted in 1968–73 (Table 3). Although the largest amount of fertilizer nitrogen tested on each crop was much larger than that in general use at that time, yields of potatoes and sugar beet on soils with less organic matter did not equal those on soils with more. In very recent years following the introduction of cultivars of wheat and barley with a high yield potential, best yields of both on Broadbalk and Hoosfield, respectively, are on soils annually treated with FYM and given extra fertilizer nitrogen (Table 4).

Table 3 *Yields of potatoes and sugar beet, spring barley, and spring wheat in 1968–73 on soils treated with fertilizers or FYM since 1846*

Crop	Manuring	Fertilizer N applied*			
		0	1	2	3
Potatoes	FYM	24.2	38.4	44.0	44.0
	PK	11.6	21.5	29.9	36.2
Sugar beet	FYM	27.4	43.5	48.6	49.6
	PK	15.8	27.0	39.0	45.6
Spring barley	FYM	4.18	5.40	5.16	5.08
	PK	1.85	3.74	4.83	4.92
Spring wheat	FYM	2.44	3.73	3.92	3.79
	PK	1.46	2.97	3.53	4.12

*N0, N1, N2, N3: 0, 72, 144, 216 $kg\,N\,ha^{-1}$ for potatoes and sugar beet; 0, 48, 96, 144 $kg\,N\,ha^{-1}$ for spring barley and spring wheat.

Table 4 *Largest annual yields of winter wheat and spring barley, $t\,ha^{-1}$, given by fertilizers and farmyard manure, Rothamsted*

Experiment and period	Crop	Treatment		
		NPK*	FYM	FYM + N†
Broadbalk 1985–90	Winter wheat grown:			
	continuously	6.69	6.17	7.92
	in rotation	8.61	7.89	9.36
Hoosfield 1988–91	Spring barley grown:			
	continuously	5.21	5.50	6.06

* Soils with FYM treatment contain about 2.5 times as much organic matter as those given fertilizers.
† N 96 $kg\,ha^{-1}$ as fertilizer.

3 Amounts of Organic Matter in Soil

The amount of humus in soil depends on (i) the amount of added organic material and its rate of decomposition; (ii) the rate of decomposition of existing humus; (iii) soil texture; (iv) climate. Soil humus usually has a carbon:nitrogen (C:N) ratio of about 10:1 despite the wide range of C:N ratios of added organic material which can be within the range 100:1 (straw) to 30:1 (leguminous crop residues). Both factors (i) and (ii) above are affected by the farming system practised, but for any one farming system on a given soil type, soil humus tends towards an equilibrium value. For example this is about 1.7% and 5.0% organic matter in permanent arable and grassland soils, respectively, at Rothamsted.

The quantity of organic matter added to soil at any one time is rarely more than a very small fraction of the humus in soil. Any addition is a welcome source of food for the ever hungry soil microbial population, the carbon content of which rarely exceeds about 5% of the total organic carbon in soil. Provided soil moisture and temperature are above the critical values below which microbial activity ceases, the microbial biomass rapidly attacks added organic material of plant or animal origin. For example, added cereal straw loses about 40% of its dry weight in 40 days after incorporation in autumn. Some residues with a narrower C:N ratio may almost totally disappear in this period. This rapid rate of decay may have implications for the quantity of low molecular weight organic molecules in the soil solution. Concentrations of these molecules may also be affected by microbial activity in the rhizosphere and by root exudates.

The concept that humus is a comparatively uniform end-product of microbial activity is well shown in Figure 1. In the Market Garden experiment at Woburn, FYM, FYM compost, sewage sludge, and sludge compost were each added to soil at 37.5 and 75 t ha^{-1} of fresh material each year for between 20 and 25 years during the period 1942–67 (for details see reference 5). These additions resulted in different amounts of humus in the eight differently treated soils. But once the applications ceased the rate of humus decay for each treatment fitted a single decay curve.[8] Presumably soluble organic molecules produced by the decay of humus could have very similar properties independent of the form in which organic matter was originally added to soil.

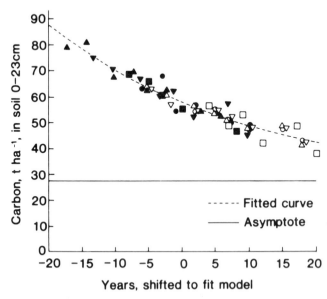

Figure 1 *Decline in soil carbon,* t ha^{-1}, *in a sandy loam, Market Garden experiment, Woburn. All treatments shifted horizontally to test whether they fitted a single decay curve. FYM, single □, double ■; sewage sludge, single △, double ▲; FYM compost, single ○, double ●; sludge compost, single ▽, double ▼*

The slow rate of change of humus in temperate soils, together with the effect of soil texture is well shown in Figure 2. On the silty clay loam at Rothamsted humus levels in soils which have been unmanured or given inorganic fertilizers since 1852 have remained essentially constant during the last 100 years. Soils given fertilizers contained about 10% more organic matter because they grew bigger crops and larger organic residues were returned to the soil each year. Soil given 35 t ha^{-1} FYM each year has still not reached its equilibrium humus content although the annual rate of gain is now very small. In 1876 the soils of Stackyard Field at Woburn had been in arable cropping for some years but nevertheless contained more humus than similarly cropped soils at Rothamsted. This was probably because the Woburn soils had a long period in grassland before the 1830s.[9] However, since 1876 the Woburn soils, which have had similar cropping and manuring to the soils on Hoosfield, have gradually lost humus and they now contain less than the Rothamsted soils do.

The magnitude of the changes in humus with changes in cropping have been discussed in detail elsewhere.[7] In one example a grassland soil with about 5.2% organic matter was ploughed. After about 20 years, 30% of the humus had been lost where a six course rotation (three cereals, two root crops, and a one year ley) was followed. Where more root crops were grown, four in six years, about 40% of the humus was lost, presumably because there was more soil cultivation for weed control and less root residues were returned. Where the soils were kept without crop and without weeds, nearly 50% of the humus had gone in the same period.

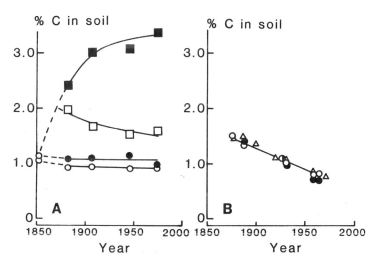

Figure 2 *Percentage carbon in soil. A. Hoosfield, Rothamsted, barley each year and treatments continuous since 1852. ○, unmanured; ●, NPK fertilizers; □, FYM 1852–71 none since; ■ FYM, 35 t ha^{-1}. B. At Woburn arable crops each year; cereals, ○, unmanured; ● NPK fertilizers; △, manured four-course rotation*

In the ley–arable experiments at Rothamsted, a sequence of three years of leys and three arable crops was compared with a six year, all-arable rotation. After 27 years the grass and grass–clover leys had done little to increase total humus content compared to that in the all-arable rotation whilst lucerne had no effect at all (Table 5). However, as already mentioned, the leys especially the lucerne, did leave a readily mineralized organic residue, probably too small to determine accurately, which did affect the response of the following arable crops to fertilizer nitrogen. On a soil initially with about 5% organic matter, continuous arable cropping caused the humus content to fall and the short leys did little to prevent the decline. On soils with little humus at the start, the short leys did little to increase humus content; it was only under permanent grass that soil organic matter content increased.[10]

4 Humus and Nitrate Leaching

One of the major concerns in the agriculture/environment debate of the late 1970s and 1980s has been the amount and source of nitrate in potable waters. Many people have considered that there was a simple direct relationship between the increasing use of nitrogen fertilizers, especially on arable crops, throughout this period and the increasing levels of nitrate observed in some water sources. Research, using a range of arable crops and ^{15}N-labelled inorganic nitrogen fertilizers, showed that, provided the amounts of fertilizer were adjusted to meet expected crop requirements and that the application was timed to satisfy crop demand for nitrogen, then very little ^{15}N-labelled nitrate remained in soil after harvest. Thus the use of inorganic nitrogen fertilizer with best farm practice had little direct effect on nitrate leaching. However, about 20–30% of the spring applied nitrogen was still in the soil but in organic combination and, especially after cereal crops, little of this new organic matter was mineralized rapidly.

Table 5 *Effect of three-year leys on % C in air dry soil, 0–23 cm*. Ley–arable experiments, Rothamsted, mean 1972 and 1975†*

	Continuous arable	Three years arable preceded by three years of:		
		Lucerne	*Grass clover*	*Grass with N*
Old grassland soil				
% C in soil	2.01	2.00	2.27	2.24
increase due to ley		−0.01	+0.26	+0.23
Old arable soil				
% C in soil	1.57	1.57	1.81	1.77
increase due to ley		0	+0.24	+0.20

* Soil sampled in autumn of third year of the ley before ploughing.
† 24th and 27th year after the start of the experiment.

Nitrate that appeared in soil after harvest came from the mineralization of humus.

Thus the problem of nitrate leaching is related to the mineralization of humus in autumn when crop demand is small. In largely arable cropping systems the long continued use of nitrogen fertilizers does increase soil organic matter a little (see, *e.g.* Figure 2) relative to the amounts in unmanured soils. Mineralization of this extra organic matter will give some additional nitrate. However, per unit area of land, the amount is likely to be small relative to the quantity of nitrate released following the ploughing of leys and the incorporation of leguminous crop residues.

5 Effects of Humus on the Movement of Organic Chemicals

A well documented effect of soil organic matter is on the mobility of some organic molecules added to soil to control weeds or pests. Here it is essential to differentiate between the effects of elemental carbon and humus. Elemental carbon, often found in soil as coal or charcoal, has also been added to soil in recent times as a result of straw burning. Usually this carbon is very finely divided and is considered to have good sorbance properties. Elemental carbon is determined when soils are analysed by dry combustion and any such carbon multiplied by the conventional factor 1.72, will be reported as organic matter. The decreased efficacy of some soil applied organic chemicals has been related to the presence of elemental carbon. Carbon in humus is still in organic combination and will have very different properties to elemental carbon.

Table 6 shows yields of field beans (*Vicia faba*) grown in an experiment where plots had been treated with inorganic fertilizers or FYM, which had resulted in soils with different humus contents, but there had never been any straw burning. Simazine was used to control weeds but, on soils low in organic matter, the simazine was not retained in the top soil. It moved downwards, as shown by bioassay in the glasshouse, and adversely affected germination, growth, and final grain yield.

Table 6 *Effect of simazine on yield ($t\,ha^{-1}$) of field beans* (Vicia faba) *at three levels of soil organic matter*

% Organic matter	Simazine added ($kg\,ha^{-1}$)	
	0	0.94
1.2	2.5	1.6
2.0	2.6	2.4
3.5	3.1	2.9

6 Effect of Humus on Metal Content of Soils and Crops

Joint research by Lancaster University and Rothamsted on cadmium in crops and soils of the Rothamsted Classical experiments has been summarized recently.[11] On soils which have received no agricultural amendments, aerial deposition of cadmium has increased the level of soil cadmium over time (Table 7). Soils with $pH_{(water)}$ above 6.5 and low in organic matter (Broadbalk, Hoosfield, Barnfield) have apparently retained none of the cadmium added in superphosphate, made from rock phosphate, since 1889. Before 1889 the superphosphate was made from bone dust which would have contained little or no cadmium. In contrast acid soils with more organic matter (Park Grass) have retained additional cadmium. Much of the extra cadmium in the soils richer in organic matter has accumulated in the near surface layers (Table 8).

Cadmium concentrations in herbage from Park Grass ranged between 102 and 152 μg Cd kg^{-1} between 1861 and 1920, increased sharply to about 250 μg kg^{-1} by the 1940s, and have remained consistently above 200 μg kg^{-1} since then. On the unmanured plot which has been limed (given $CaCO_3$) once every four years since 1903, soil pH is now higher

Table 7 *Changes in soil cadmium on untreated plots and those given superphosphate in Rothamsted long-term experiments*

Experiment	Period	Approximate average yearly increase			
		Soils without P		Soils with P	
		(μg kg^{-1})	(g ha^{-1})	(μg kg^{-1})	(g ha^{-1})
Broadbalk	1881–1983	1.0	2.9	0.9	2.6
Hoosfield	1882–1982	1.5	3.9	1.6	4.1
Barnfield	1870–1983	0.9	2.3	1.2	3.1
Mean		1.1	3.0	1.2	3.3
Park Grass	1876–1976	1.1	2.9	2.7	7.2

Table 8 *Cadmium* (mg kg^{-1}) *in soil from different depths, Park Grass, Rothamsted*

Soil layer (cm)	$-P$*	$+P$	Increase due to superphosphate
0–7.5	0.21	0.40	0.19
7.5–15.0	0.15	0.28	0.13
15.0–22.5	0.12	0.18	0.06
22.5–30.0	0.08	0.12	0.04
30.0–37.5	0.04	0.08	0.04
37.5–45.0	0.02	0.03	0.01

*$-P$, $+P$ without and with superphosphate.

(pH 6.4) than on the unlimed plot (pH 5.2), and herbage cadmium concentrations have increased only a little. This suggests that some of the cadmium could have been on the 'outside' of the herbage directly from aerial deposition whilst some was on the 'inside' from root uptake. On the limed soils, any effect of decreasing acidity on lowering cadmium uptake by roots was offset by increased aerial deposition so that concentrations changed little over time. On acid soils root uptake may have been enhanced which, together with increased aerial deposition, caused concentrations to increase. Cereal grain cadmium concentrations were little affected by increased cadmium concentrations in surface soil.[11]

Figure 3 shows the concentration of herbage lead since the late 1950s in another experiment. Lead concentrations have declined throughout the period and were related to a decline in aerial lead. In this experiment herbage grown on soils with and without FYM was analysed. The FYM caused a small increase in soil organic matter (from 5.0% to 6.4%) but this had no effect on herbage lead. This suggests that much of the lead was from aerial deposition on to the outside of the plant. Soil lead burden has increased where 35 t ha^{-1} FYM has been applied each year in the Classical experiments[12] but the amounts were small. Probably much of the animals dietary intake of lead was excreted firmly complexed with the organic fraction and then retained in soil.

This evidence for cadmium and lead suggests that organic matter can play a role in retaining both metals in surface soil. It also suggests that root uptake of both metals will depend on their speciation in soil and for cadmium, uptake can be manipulated by altering soil pH even in the presence of organic matter.

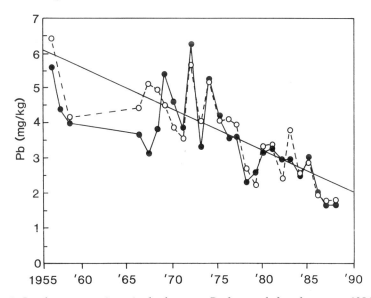

Figure 3 *Lead concentrations in herbage at Rothamsted for the years 1956–1988. Soils given,* ●, *NPK fertilizers;* ○, *FYM*

7 Humus and the Movement of Phosphorus

There is concern currently about the sources of phosphorus in inland and coastal waters because of its role in the development of algal blooms when both phosphorus and nitrogen (as nitrate) are above critical thresholds. Much of the nitrate comes from agricultural land, mainly from the mineralization of organic matter. Appreciable amounts of phosphorus can come from point sources (like sewage treatment works) and there is discussion about losses from agricultural land. Studies at Rothamsted on the movement of phosphorus in agricultural soils have been summarized recently.[13]

Table 9 shows the total P content at different depths in soils in one experiment where root crops were grown with NPK fertilizers $(33 \text{ kg P ha}^{-1})$ and FYM $(35 \text{ t ha}^{-1}$, containing about $40 \text{ kg P ha}^{-1})$ since 1843. The plough layer is now 23 cm deep and there has been considerable enrichment with phosphorus where both fertilizers and FYM were given. Where superphosphate was applied there was a small increase in the phosphorus concentration in the 23–30 cm layer which could be due to a limited amount of leaching but is more likely to be the result of occasional ploughing just deeper than 23 cm. There has been no phosphorus enrichment of the soil below 30 cm. Where FYM was applied there has been appreciable enrichment of the 23–30 cm and 30–46 cm horizons. These data are summarized and compared with those from other long-term experiments in Table 10. The permanent grassland experiment has not been ploughed for perhaps the last 300 years and unmanured plots and those given superphosphate $(33 \text{ kg P ha}^{-1})$ have had the same treatment each year since 1856. In the 0–46 cm layer of the unmanured soil there is the same amount of total P as in the comparable soil of the arable experiment but the distribution is different. Where superphosphate has been applied there has been considerable phosphorus enrichment of both the 23–30 and 30–46 cm layers. When averaged over all treatments for the grassland soil, the increase in total P (670 mg kg^{-1}) below 23 cm was about

Table 9 *Total phosphorus, mg P kg^{-1}, content of soil in 1958 at various depths where superphosphate and farmyard manure had been applied since 1846, Barnfield, Rothamsted*

Depth (cm)	*Treatment*				
	No P No K	*P No K*	*PK*	*FYM*	*FYM + P*
0–23	770	1350	1295	1376	1971
23–30	464	541	524	649	781
30–46	413	446	448	525	579
46–53	401	396	397	442	411

Each value is the mean of five nitrogen treatments.

80% of the average increase $(820 \, mg \, P \, kg^{-1})$ in the top 23 cm.[14] All plots of vegetable crops grown on the sandy loam in the Market Garden experiment at Woburn received superphosphate whilst some were given FYM. Where FYM was given subsoils have been enriched with phosphorus (Table 10).

Phosphorus enrichment of the subsoils has occurred, therefore, where FYM but not superphosphate was applied to mineral soils growing arable or vegetable crops and where superphosphate was applied to soil rich in organic matter. This suggests that P may be transported as water-soluble, low molecular weight organic phosphorus molecules present in FYM or, in the case of grassland soils, released on the death of roots. When these soils were extracted with different reagents the phosphorus in surface soils from which phosphorus has leached had an enhanced solubility in $0.01 \, M \, CaCl_2$ solution. This solution has a similar ionic strength to that of the soil solution. Any organic phosphorus compounds extracted by this reagent would be readily hydrolysed when the solution was acidified to measure its phosphorus content and would be determined as inorganic phosphorus.

Table 11 shows the total, bicarbonate-soluble and $CaCl_2$-soluble P in control, FYM and superphosphate treated soils in a number of long-term experiments. For the four experiments growing arable crops on mineral soils at Rothamsted the increase in total and bicarbonate soluble P from FYM and superphosphate was essentially the same but the increase in $CaCl_2$-soluble P was between two and five times greater in FYM-treated soils. On Barnfield, where both FYM and superphosphate were applied together, the presence of FYM enhanced the $CaCl_2$-soluble P in the combined treatment more than the sum of the separate effects of FYM and superphosphate. This effect of FYM occurs only when it is added regularly; it had disappeared on the Exhaustion Land by 1974 where the last application of FYM was in 1901 (Table 11) even though the FYM-treated soil still contained about 25% more total organic matter than the fertilizer-treated soil. Table 11 also shows that at Woburn, although sewage sludge-treated soils contained nearly three times as much extra total phosphorus as FYM-treated soils, they had much less extra bicarbonate-soluble P and $CaCl_2$-soluble P. The sewage sludge used between 1942 and 1961 had been anaerobically digested and lagoon dried so it is probable that much soluble P, both inorganic and readily mineralized organic P, had been removed in the treatment works and discharged to the river.

Some of the phosphorus from superphosphate applied to grassland soils has moved down through the soil horizons by leaching. However, in this situation the extra $CaCl_2$-soluble P decreased more rapidly with depth than did bicarbonate-soluble or total P.[13] This suggests that if low molecular weight organic molecules were transported downwards they were rapidly mineralized and the inorganic phosphorus was adsorbed on to sites in the subsoil. Thus organic matter, directly or indirectly appears to play a role in the movement of phosphorus downwards through the profile of agricultural soils.

Table 10 *Total P*, $mg\,kg^{-1}$, *in soil at different depths where superphosphate or farmyard manure were applied to surface soils at pH 6.5*

Soil depth (cm)	Soil type and P treatment*						
	Silty clay loam					Sandy loam	
	Arable crops			Permanent grassland		Vegetable crops	
	None	P	FYM	None	P	P	FYM
0–23	780	1295	1375	575	1425	1120	1780
23–30	465	525	650	555	785	–	–
30–46	415	450	525	500	600	850	960
Below 46	400	395	440	–	–	790	860

* P superphosphate, FYM farmyard manure.

Table 11 *Total, $NaHCO_3$-soluble and $CaCl_2$-soluble P in surface soils, 0–23 cm, from various long-term experiments at Rothamsted and Woburn*

Experiment and year sampled	Treatment	Total P $(mg\,kg^{-1})$*	P soluble in 0.5 M $NaHCO_3$ $(mg\,kg^{-1})$*	P soluble in 0.01 M $CaCl_2$ $(g\,mol^{-1} \times 10^6)$*
Barnfield 1958	Control	670	18	0.5
	Superphosphate (P)	1215 (545)	69 (51)	3.0 (2.5)
	FYM	1265 (595)	86 (68)	12.8 (12.3)
	FYM plus P	1875 (1205)	145 (127)	22.3 (21.8)
Broadbalk 1966	Control	580†	8	0.2
	Superphosphate	1080 (500)	81 (73)	6.6 (6.4)
	FYM	1215 (635)	97 (89)	19.5 (19.3)
Hoosfield 1965	Control	630	6	0.3
	Superphosphate	1175 (545)	103 (97)	14.4 (14.1)
	FYM	1340 (710)	102 (96)	25.4 (25.1)
Exhaustion Land 1903	Control	530	8	0.2
	Superphosphate	885 (355)	65 (57)	5.6 (5.4)
	FYM	860 (330)	66 (58)	9.6 (9.4)
1974	Control	480	2	0.1
	Residues of:			
	Superphosphate	595 (115)	10 (8)	0.2 (0.1)
	FYM	630 (150)	12 (10)	0.3 (0.2)
Woburn 1961	Control	1120	94	23
	FYM	1780 (660)	176 (82)	98 (75)
	Sewage sludge	3000 (1880)	151 (57)	43 (20)

* Figures in parenthesis are increases over the control.
† 1944 samples.

8 Conclusions

Soil organic matter or humus plays a crucial role in soil fertility. Humus is the end-product of the microbial decomposition of vegetable or animal organic matter deposited on or in soil. This process is essential for the well-being of soil. It results in a product of relatively uniform C:N ratio which has important physico-chemical properties. The quantity of soil humus depends on the farming system practised, on soil type and climate. For any one husbandry system heavier textured, clayey soils contain more humus than light textured, sandy soils. The humus content for any combination of soil and farming system tends towards an equilibrium value and so there is no universal critical value for humus content. In temperate climates the rate of change is slow and long-term experiments are usually needed to estimate the effects of humus on yields of arable crops. Increased yields on soils with more humus have been measured in recent years especially for spring sown crops with a high yield potential. A number of factors, including better soil structure, water holding capacity and availability of nutrients can contribute to these beneficial effects. Crop quality can be affected through interactions effecting the availability of both nutrients and pollutants.

Besides benefits, there can be problems. When farmed in accordance with good farm practice, mineralization of organic matter is the major source of nitrate at risk to loss by leaching. Humus can retain and affect the availability of some pollutants like cadmium and low molecular weight organic molecules appear to be implicated in the movement of phosphorus downwards through the soil profile. But there is no way in which we could farm without humus.

The many complex interactions between soil and mineral particles, added organic matter, humus, microbial activity, and transport processes are, as yet, not fully understood. But such understanding is crucial in seeking ways to minimize the risk of pollutants moving through soil to water used for human consumption. A knowledge of the role of humus in transport processes is only one part of the story, but a very important one because soil organic matter is crucial to crop production.

References

1. J. Liebig, 'Organic Chemistry in its Application to Agriculture and Physiology', Taylor and Walton, London, 1840, p. 200.
2. J.B. Lawes, *J. R. Agric. Soc.*, 1847, **8**, 226.
3. E.W. Russell, *Philos. Trans. R. Soc. London, B*, 1977, **281**, 209.
4. MAFF, 'Modern Farming and the Soil', Report of the Agricultural Advisory Council on Soil Structure and Soil Fertility (The Strutt Report), MAFF, HMSO, London, 1970.
5. A.E. Johnston and R.W.M. Wedderburn, Rothamsted Expt. Stn. Report for 1974, Part 2, 1975, p. 79.

6. G.E.G. Mattingly, M. Chater and P.R. Poulton, Rothamsted Expt. Stn. Report for 1973, Part 2, 1974, p. 134.
7. A.E. Johnston, *Soil Use Manage.*, 1986, **2**, 97.
8. A.E. Johnston, S.P. McGrath, P.R. Poulton and P.W. Lane, 'Nitrogen in Organic Wastes Applied to Soils', Academic Press, London, 1989, p. 126.
9. A.E. Johnston, 'Woburn Experimental Farm: a Hundred Years of Agricultural Research Devoted to Improving the Productivity of Light Land', Lawes Agricultural Trust, Harpenden, 1977.
10. A.E. Johnston, Rothamsted Expt. Stn. Report for 1972, Part 2, 1973, p. 131.
11. A.E. Johnston and K.C. Jones, 'Phosphate Fertilizers and the Environment', International Fertilizer Development Centre, Muscle Shoals, Alabama, USA 1992, p. 255.
12. K.C. Jones, C.J. Symon and A.E. Johnston, *Sci. Total Environ.*, 1987, **61**, 131.
13. A.E. Johnston and P.R. Poulton, 'Phosphate Fertilizers and the Environment', International Fertilizer Development Centre, Muscle Shoals, Alabama, USA 1992, p. 45.
14. A.E. Johnston, 'Agriculture and Water Quality', MAFF Tech. Bull. 32, HMSO, London, 1976, p. 111.

2

Concentration and Composition of Dissolved Organic Carbon in Soils, Streams, and Groundwaters

By Ronald L. Malcolm

US GEOLOGICAL SURVEY, BOX 25046, MS 408, DENVER FEDERAL CENTER, DENVER, COLORADO 80225, USA

Introduction

Organic water quality studies have become common only in the last two decades with the advent of environmental concerns. Regular studies are now made of specific organic compound analyses for pesticides and herbicides, which may be harmful to natural fauna and flora; for trihalomethanes and other chlorinated compounds, which may be harmful to human health; and for industrial chemicals, which may damage the natural ecology. While studies of these compounds have become synonymous with organic water quality, in fact, the total of such compounds represents less that 1–2% of the organic constituents in water. It has been slowly recognized that the major organic constituents in water are the natural organic solutes which are important in all aspects of water quality and do more than cause interference with the analyses of specific pollutant compounds.

The first attempt at general organic water quality analysis was the determination of total organic carbon (TOC). Due to the variable concentration of particulate carbon in water samples, and the enormous differences in composition and reactivity of organic constituents in the dissolved phase as compared to that of the suspended phase, the TOC parameter was an uninterpretable organic water quality parameter and has been discarded as obsolete by most scientists. The dissolved organic carbon (DOC) parameter was introduced by Malcolm et al. in 1973,[1] and has become the most widely used and the most easily interpretable parameter of general organic water quality. The development of liquid-phase (DOC) carbon analysers, of membrane filters to efficiently separate dissolved and particulate phases, and of resin technology to isolate organic constituents of the dissolved phase has facilitated the use and proliferation of research on the DOC in water.

The major purpose of this paper is to review the progress to date in DOC water research relative to the DOC concentration and the composition of DOC in various natural environments. The many important aspects of DOC in environmental and geochemical processes have been well documented in many other publications and will not be emphasized in this paper.

2 Methodology and Terminology for Dissolved Organic Carbon Fractions

A brief discussion of DOC methodology and terminology for DOC fractions is necessary for easy understanding of this paper by scientists who are generally unfamiliar with DOC research.

Almost all DOC data are understood to be qualified as non-volatile or non-purgeable DOC. In almost all surface waters the volatile DOC components have been found to be less that 1% of the DOC.[2] The volatile component in groundwaters has yet to be determined due to the special pressurized sampling techniques required, but it is assumed to be a small component of the total DOC. This assumption is obviously invalid for groundwaters contaminated with volatile hydrocarbons.

Water samples for DOC analysis should be pressure filtered in a stainless-steel filter barrel (such as a Gelman model 4280)* fitted with a 0.45 μm silver membrane filter, or a plastic membrane filter free from organic wetting agents which will seriously contaminate the sample. The filtered water should be collected into a glass vial which has been baked at 450 °C to make it organic free, and the filled vial capped with a PTFE-lined cap, and then the sample vial should be stored on ice at 2–4 °C in the dark until analysis. Acidification or addition of sample preservatives results in losses of DOC from the sample or causes potential problems during DOC analysis. Sample freezing is undesirable because most samples flocculate upon thawing, causing serious sub-sampling problems.

A number of DOC instrumental analysers may be used for DOC analysis, but DOC sample precision must be ± 0.1 mg carbon l^{-1}(C l^{-1}) or better for correct data interpretation, especially in determining DOC fractionation analysis. Direct analysis of DOC after acidification, then purging the inorganic carbon, is recommended rather than a difference determination of total carbon minus inorganic carbon. DOC analysis by dichromate wet oxidation is unacceptable due to incomplete oxidation of organic constituents.

The DOC fractionation procedure on XAD-8 resin and cation exchange resins is the most common and acceptable method of separating DOC constituents into chemical classes.[3] The six fractions are hydrophobic acids, hydrophobic bases, hydrophobic neutrals, hydrophilic acids, hydrophilic

---

*The use of the Gelman trade name in this report is for identification purposes only, and does not constitute endorsement by the US Geological Survey.

bases, and hydrophilic neutrals. In most waters, the hydrophobic and hydrophilic bases are such a low percentage (1–4%) of the total DOC that these two fractions can be ignored in the isolation procedure. By flowing the effluent of the XAD-8 column into a second column of XAD-4 resin, approximately 90% or more of the hydrophilic acid fraction from XAD-8 can be isolated from the XAD-4 resin column. By this tandem resin method 80 to 85% of the DOC can be concentrated and the four fractions can be isolated.[4] The hydrophilic neutral fractions can be analysed for specific compound analysis according to the desires of the investigator. The hydrophobic acid fraction can also be divided into fulvic acids and humic acids if desired.

The word hydrophobic is a generic relative term. All the DOC is hydrophilic or dissolved in water. The individual DOC components have differing molar solubilities or hydrophilicity in water. The classification term hydrophobic is relative to sorption of solutes onto the XAD resins at pH 2. It has been determined that all fulvic and humic acids in water have a k', column capacity factor, of 50 or greater and are sorbed out onto the XAD-8 resin at a pH of 2. They are classified as hydrophobic acids because they are sorbed or hydrophobic relative to XAD-8 resin at pH 2. At pH 2 the hydrophobic acids are protonated to be neutral or nearly neutral species. The hydrophobic neutrals are true neutral constituents which are neutral at all pHs and would sorb onto XAD-8 regardless of pH adjustment of the water sample.

The hydrophilic acids, according to the DOC fractionation scheme, are hydrophilic relative to XAD-8 and pass through the XAD-8 column at pH 2 as unsorbed hydrophilic solutes. The majority (approximately 90%) of this hydrophilic acid fraction relative to XAD-8 resin becomes hydrophobic relative to XAD-4 resin at pH 2, is sorbed onto the XAD-4 resin at a solute k' of 50, and is then eluted from XAD-4 with dilute base. This acid fraction is called 'XAD-4 acids'.

3 Dissolved Organic Carbon in Streams

The DOC of streams has been the most extensively studied of all natural environments. The DOC of uncoloured streams ranges from 2 to 10 mg Cl^{-1}. Uncoloured polluted streams usually are in the same DOC range, but some may be higher than 10 mg Cl^{-1}. Organically coloured streams (often called humic waters) have a broad range of DOC concentrations; they may be as low as 3 mg Cl^{-1} to in excess of 50 mg Cl^{-1}.

The ratio of hydrophobic:hydrophilic organic constituents in most uncoloured surface waters is 60:40. The average DOC distribution is 48% fulvic acids, 4–5% humic acids, 6–8% hydrophobic neutrals, 25% hydrophilic acids, 2–4% basic compounds, and 8–10% hydrophilic neutrals. The ratio of fulvic acids to humic acids in most uncoloured streams is 10:1. The hydrophobic:hydrophilic ratio of humic waters may increase to 70:30 or higher due to the higher concentrations of humic acids. Hydrophobic acids

Concentration and Composition of Dissolved Organic Carbon

(fulvic plus humic acids) often account for 60–65% of the DOC with humic acids accounting for 15–20% of the DOC. The resulting fulvic:humic ratio in such waters is between 5:1 and 3:1.

The cross-polarization magic angle spinning (CPMAS) ^{13}C NMR spectra of four typical fractions isolated from surface water (fulvic acids, humic acids, XAD-4 acids, and hydrophobic neutrals) are shown in Figure 1. These four fractions represent 85% of the DOC from Lake Skjervatjern, a small lake near Forde, Norway. The detailed ^{13}C NMR spectral interpretations are discussed by Malcolm.[5] These spectra show large compositional differences among these fractions. The hydrophobic neutral fraction is highly aliphatic (1–110 ppm), with low aromaticity (110–162 ppm) and low

LAKE SKJERVATJERN DOC FRACTIONS

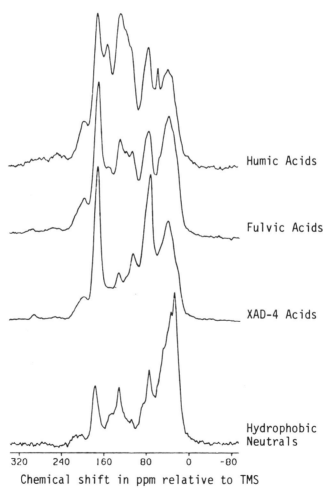

Figure 1 *CPMAS ^{13}C NMR spectra of Lake Skjervatjern DOC fractions*

carboxyl content (162–195 ppm). The XAD-4 acids are moderately high in aliphaticity, low in aromaticity, but are very high in carboxyl content. The humic acids are moderate in aliphatic character, high in aromatic content, and also high in carboxyl content. The humic acids spectrum also exhibits methoxyl peaks (50–60 ppm) and phenolic peaks (145–162 ppm). The fulvic acids are high in aliphatic content, moderate in aromatic content, and high in carboxyl content. Fulvic acids are also low in methoxyl and phenolic contents.

4 Dissolved Organic Carbon in Groundwaters

The mean DOC concentration in more than 100 deep-seated groundwaters throughout the United States was found to be 1 mg Cl^{-1} with a range from 0.2 to 15 mg Cl^{-1}.[6] Shallow groundwater aquifers that are subject to high infiltration rates and potential contamination from the land surface often have higher DOC concentrations, between 2 and 5 mg Cl^{-1}. No extensive survey of European groundwaters for DOC content is known by the author, but the limited published data suggest the DOC of European groundwaters is in the same concentration range of groundwaters as in the United States.

In a few localized groundwater environments in the western United States, highly coloured waters called trona waters, with a colour intensity of motor oil, have DOC concentrations in excess of 1000 mg Cl^{-1}. These waters are believed to have formed in sub-surface environments where highly alkaline waters have solubilized lignite and other weathered coal deposits. The organic constituents of trona waters are also highly unusual in composition because they contain 70–75% carbon and are still water soluble.

Groundwaters are commonly characterized by a hydrophobic:hydrophilic ratio of 40:60 which is the converse for normal surface waters. The distribution of organic solutes is 50% hydrophilic acids, 15–20% fulvic acids, 2–5% humic acids, essentially no basic constituents, 4–5% hydrophobic neutrals, and 20% hydrophilic neutrals.

Groundwaters contain much lower DOC concentrations than near-surface soil interstitial waters or interflow waters during rainfall. This is due to the preferential sorption of hydrophobic DOC constituents as compared to hydrophilic DOC constituents onto sediment surfaces as percolating waters move downward in soils and surficial sediments, and to the long-term *in situ* decomposition of humic substances by organisms within the aquifer. By using computer mixing models,[7] the relative contributions of these two water sources to stream discharge can be estimated. It is suggested that similar mixing models based on hydrophobic:hydrophilic organic constituent differences between groundwaters and run-off surface waters could be accomplished to estimate groundwater flow contributions to surface streams. The agreement of the two models should confirm the utility of these types of hydrologic models.

The composition of the hydrophobic neutral fraction, the hydrophilic neutral fraction, and the hydrophilic acid fraction of groundwater DOC has yet to be reported. The application of the tandem resin isolation procedure to isolated XAD-4 acids has also not yet been applied to groundwaters. Fulvic and humic acids have been isolated from a few groundwaters. The compositions of the respective isolated groundwater fulvic and humic acids are very different from their respective counterparts in surface waters or soils.[8] Groundwater humic substances which are predominantly fulvic acids are higher in carbon content (62 *versus* 54%), higher in hydrogen content (6 *versus* 4%), lower in oxygen content (26 *versus* 39%), and lower in nitrogen content (0.4 *versus* 1.2%) than for surface water humic substances. The CPMAS ^{13}C NMR spectra for groundwater fulvic acids (as shown in Figure 2) are much lower in C—O and phenolic carbon functionalities than surface-water fulvic acids due to the labile nature of these functionalities. Lastly, groundwater fulvic acids are very light tan in colour with a colour intensity per carbon atom approximately 10% of surface-water fulvic acid.

5 Dissolved Organic Carbon of Soil Interstitial Waters

The DOC aspects of soil interstitial water is one of the least studied of all DOC environments. The few studies that have measured DOC concentrations in soil interstitial waters have been in Spodosols or closely related soils in acid-rain environments.[9-12] DOC concentrations between 5 to 15 mg Cl[-1] have been reported with the highest concentrations near the soil surface. A decrease in hydrophobic constituents and an inverse in hydrophilic acids with depth in soil has been reported.[13]

Other studies of soil interstitial waters have been conducted by the author by determining the concentration and composition of the water-extractable or leachable DOC from selected soil horizons. The water leached from these soils by distilled water at a rate and periodicity simulating natural rainfall are filtered and then processed as a surface-water sample for the isolation of DOC constituents by the tandem resin isolation procedure.[4,14] The soils' orders leached to date have been Spodosol, Inceptisol, Mollisols, and Histosols. DOC concentrations of water extracts of soils range from 10 to 1000 mg Cl[-1]. DOC concentrations of some continuously extracted soils are maintained as high as 50 mg Cl[-1] with frequent leaching.

A generalized DOC distribution according to compound class cannot be yet stated for soil interstitial waters due to the sparsity of data, but certain trends appear to be evident. The hydrophobic:hydrophilic separation in surface-soil horizons appears to be very dependent on drainage class. The more poorly drained the soil, and with the prevalence of saturated conditions, the more dominant are hydrophilic acids and hydrophilic neutrals, thus leading to a higher hydrophobic:hydrophilic ratio in the

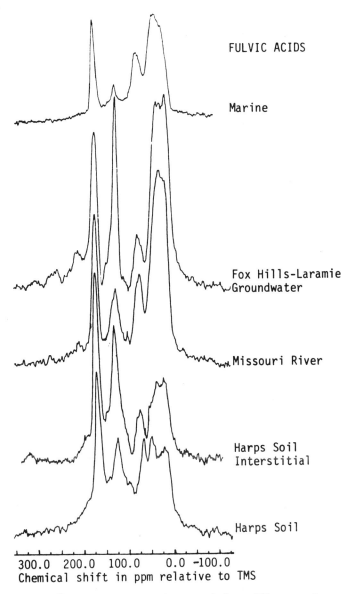

FULVIC ACIDS

Marine

Fox Hills-Laramie
Groundwater

Missouri River

Harps Soil
Interstitial

Harps Soil

300.0 200.0 100.0   0.0 -100.0
Chemical shift in ppm relative to TMS

Figure 2 *CPMAS ^{13}C NMR spectra of fulvic acids from different environments*

order of 40:60. Well-drained surface soils tend to exhibit hydrophobic:hy-drophilic ratios of near 60:40. Just as in all waters, the dominant organic species are organic acids.

The concentration of fulvic acids in most soil interstitial waters is much higher than that of humic acids, resulting in a fulvic acids:humic acids ratio of between 5:1 and 10:1. Exceptions to this ratio are exhibited by the Harps soil mollic epipedon interstitial water with a fulvic:humic ratio of 2:1

and on the other extreme, the Kachemak B_{hs} horizon interstitial water which only has trace amounts of humic acid. These findings strongly suggest that humic acids are more hydrophobic than fulvic acids; therefore, humic acids are preferentially sorbed in surface mineral horizons. This relative effect would also be expressed for fulvic acids and hydrophilic acids (XAD-4 acids). The fulvic acids are more hydrophobic than the XAD-4 acids; therefore, the XAD-4 acids are more mobile and should be more predominant in the lower soil horizons as compared to fulvic acids.

The fulvic acid:humic acid ratio in most uncoloured surface waters is 10:1. The fulvic acid:humic acid ratio in most bulk soils except Spodosols is approximately 1:3. The ratio of 5:1 or slightly greater indicates that soil interstitial waters are intermediate in fulvic:humic ratio between bulk soil and surface waters. This finding for soil interstitial waters is expected, indicating that the more soluble fulvic acids are leached from soils by meteoric waters preferentially to the less water-soluble humic acids.[15]

The compositions of fulvic and humic acids in soil interstitial waters are commonly different from their respective counterparts in the bulk soil. This is not surprising when one considers that fulvic and humic acids are an extensive mixture of similar compounds. It is expected that the more water-soluble components of each mixture would be preferentially leached by percolating waters through the soil. As shown in Figure 2, the ^{13}C NMR spectrum of Harp soil interstitial fulvic acid is very different from the bulk soil fulvic acid. The soil interstitial fulvic acid is surprisingly higher in aromaticity than soil fulvic acid, has a lower carbon content (52 *versus* 55%), a higher oxygen content (41 *versus* 38%), and a lower nitrogen content (0.5 *versus* 1.0%). Such findings make it highly questionable whether soil fulvic acids or soil fulvic acid fractions extracted from bulk soils, can be used as representative of the, soils interstitial water fulvic acids in pedogenic research.

XAD-4 acid fractions of soil interstitial waters are very similar to XAD-4 acid fractions from bulk soil extractions, or to those found in surface waters as shown by the CPMAS ^{13}C NMR spectra of Figure 1. The XAD-4 acids are highly aliphatic and of very low aromaticity. The high anomeric carbon content (100–110 ppm) associated with the high content of C—O carbons (50–95 ppm), and the high carboxyl content (162–195 ppm) suggest that this fraction may be high in sugar acids which are highly branched and hydroxylated. The XAD-4 acids are very high in carboxyl acidity (6.5 or more $meq\,g^{-1}$), lower in carbon content (46–48%), and are very high in oxygen content (40–44%). All these characteristics combine to give the XAD-4 acids a density of 1.7 or greater which is approximately 10% higher than for humic or fulvic acids.

The relative importance of high concentrations of hydrophilic acids (XAD-4 acids) in soil interstitial waters (one third or more of the DOC) in pedogenic processes has been overlooked in previous soil organic research. The XAD-4 acids may be as important as, or more important than, fulvic acids in metal complexation and metal elluviation. The XAD-4 acids may also be a major organic carbon constituent leached from soil which may be a precursor of aquatic fulvic or humic acids.

The direct leaching of soil fulvic or humic acids from soils into surface waters has been determined to be minimal in the United States because the composition of surface water humic substances is very different from those in bulk soils or soil interstitial waters.[16] This situation may be greatly different in stream waters in upland peat areas in the northern climates of Europe, Asia, or Canada. The cooler climates, the moderately higher rainfall, and the ease of direct leaching of peaty constituents from the soils in these regions into streams may be the major reason that colour is observed in many European waters at much lower DOC concentrations than in the United States. The colour intensity per carbon atom for soil humic substances in the United States is approximately 10 times higher than for stream waters humic substances. Additional evidence that direct leaching of soil humic substances may give a greater contribution to stream humic substances than found in the United States is the CPMAS ^{13}C NMR spectral fingerprint of peats in the ^{13}C NMR spectra of stream humic substances in headwater streams from upland peats in England. These relationships are being studied in detail by Hayes and co-workers (see Chapter 3).

6 Marine Dissolved Organic Carbon

The DOC of marine waters has been frequently reported to be between 0.5 and 3.0 mg Cl^{-1}.[17-20] During the past two to three years there has been considerable debate concerning the accuracy of marine DOC data and of the methods of marine DOC determination. Sugimura and Suzuki[20] claim that most reported values of marine DOC, as determined by wet chemical methods, are as much as 50% low because of poor sampling and preservation techniques, and the incomplete conversion of organic carbon to CO_2. In a recent article by Aiken,[21] the chloride concentrations above 0.02 M are alleged to interfere with complete conversion of organic carbon constituents to CO_2. Such effects would be expected to be accentuated with marine waters due to the naturally high background chloride concentrations.

The most common method of determining marine DOC has been the wet oxidation of DOC with persulfate developed by Menzel and Vicaro.[22] This method was developed into a commercially available apparatus, called the Oceanographic International DOC Instrument. The author tested this method for complete oxidation of several organic standards including glucose and humic acids to CO_2 during the early 1970s. The method was found to completely oxidize organic standards to CO_2 and to be in agreement with freshwater DOC values as determined by Beckman and Dohrman* instrumental values. Because the author was interested only in

*The use of Beckman and Dohrman trade names in this report is for identification only, and does not constitute endorsement by the US Geological Survey.

the determination of freshwater DOC, DOC conversion was only tested at low chloride concentrations. DOC analysis of freshwater samples by this method appears to be accurate. It is only at high chloride concentrations exceeding 0.23 M, or approaching seawater concentrations, that chloride may interfere with quantitative conversion of DOC constituents to CO_2. Incomplete conversion appears to be a greater problem for high molecular weight DOC constituents than for the lower molecular weight constituents.[20]

The distribution of DOC in marine waters among chemical classes is poorly established due to the sparsity of experiments, particularly when it has been attempted to fractionate marine DOC by the XAD resin techniques used for fresh waters, and because of the problems discussed previously with accurate analysis for the whole marine DOC or its fractions. Marine fulvic acids have been isolated, and their CPMAS ^{13}C NMR spectra, as compared to surface-water fulvic acids, are shown in Figure 2. Marine fulvic acids are very different from surface-water fulvic acids in that they are more highly aliphatic than surface-water fulvic acids and marine fulvic acids are exceptionally low in aromaticity.

Some general statements concerning marine DOC are that: (i) the concentration range in marine waters is relatively narrow ($0.5–3$ mg Cl^{-1}) as compared to a broad range of concentration for surface waters ($2–100$ mg Cl^{-1}); (ii) essentially all the humic substances are as fulvic acid; (iii) the concentration of humic acid is almost undetectable; (iv) the concentration of particulate organic carbon is very low compared to DOC; (v) DOC concentrations decrease with depth from the surface euphotic zone; and (vi) DOC concentrations are much higher in near-shore environments than in the deep ocean.

7 Conclusion

Many aspects of the DOC in surface waters are known. The DOC of uncoloured fresh surface waters commonly ranges from 2 to 8 mg Cl^{-1}. The DOC distribution usually is 50% fulvic acids, 5% humic acids, 25% low molecular weight acids, 7% hydrophobic neutrals, 8% hydrophilic neutrals, and 4% or less as basic compounds. Greater than 80% of the DOC can be isolated by tandem resin techniques, XAD-8 resin first, then XAD-4 resin. The DOC concentration of coloured surface waters (humic waters) may be 50 mg Cl^{-1} or greater. In such humic waters, the humic acid distribution may be 10–15% of the DOC. The average DOC value for groundwater is near 1 mg Cl^{-1} with a few outliers in the range 0.2–15 mg Cl^{-1}. Low molecular weight acids dominate at approximately 50% of the DOC, fulvic acids are 20–25% of the DOC, and hydrophilic and hydrophobic neutrals account for the remainder of the DOC. The DOC distribution in marine waters is similar in organic compound classes to groundwater. The DOC values vary in a narrow concentration range from 1 to 3 mg Cl^{-1}.

DOC aspects of soil interstitial waters are poorly established and poorly understood. DOC values range from 5 to 1000 mg C l^{-1} with highest values near the surface of the soil. The distribution of DOC fractions varies greatly with drainage class, but acidic fractions predominate. In many soil interstitial waters, the fulvic and humic acids are very different from their respective fractions in bulk soil or adjacent streams. The importance of low molecular weight acids in pedogenic processes has been largely overlooked.

In most streams the fulvic acid and humic acid components are distinctly different in composition from their respective fractions in soils, ground-waters, or marine waters. An exception to this finding has been found in some streams adjacent to peat soils in Great Britain. In these the humic substances retain a fingerprint of their peat soil origin.

Fulvic acid and humic acid components of DOC have been the most extensively studied. These components are often incorrectly equated to be synonymous with DOC, yet they represent only approximately 50% of the DOC. The hydrophilic neutral fraction and the hydrophobic neutral fraction of DOC in all environments have been least characterized. More extensive research on these fractions is needed, especially on the hydrophilic neutral fraction which contains 75% or more of the DOC saccharide moieties.

References

1. R.L. Malcolm, J.A. Leenheer, P.W. McKinley and L.A. Eccles, in 'Methods for Analysis of Organic Substances in Water', eds. D.F. Goerlity and E. Brown, US Geological Survey Techniques for Water-Resources Investigations, 1973, 5, A3, p.34.
2. L.B. Barber, Personal Communication, US Geological Survey, Boulder, Colorado 80303.
3. J. A. Leenheer, *Environ. Sci. Technol.*, 1981, **15**, 578.
4. R.L. Malcolm and P. MacCarthy, *Environ. Int.*, 1993, **18**, 597.
5. R.L. Malcolm, *Environ. Int.*, 1993, **18**, 609.
6. J.A. Leenheer, R.L. Malcolm, P.W. McKinley and L.A. Eccles, *J. Res.: US Geological Survey*, 1974, **2**, 361.
7. N. Christophersen, C. Neal, R.P. Hooper, R.D. Vogt and S. Andersen, *J. Hydrol.*, 1990, **116**, 300.
8. R.L. Malcolm, in 'Humic Substances II. In Search of Structure,' eds. M.H.B. Hayes, P. MacCarthy, R.L. Malcolm and R.S. Swift, Wiley, Chichester, 1989, Ch. 12, p.339.
9. R.L. Malcolm and R.J. McCracken, *Soil Sci. Soc. Am. Proc.*, 1968, **32**, 834.
10. H.J. Dawson, F.C. Ugolini, B.F. Hrutfiord and J. Zachara, *Soil Sci.*, 1978, **126**, 290.
11. W.H. McDowell and T. Wood, *Soil. Sci.*, 1984, **137**, 23.
12. C.S. Cronan and G.R. Aiken, *Geochim. Cosmochim. Acta*, 1985, **49**, 1697.
13. K.B. Easthouse, J. Mulder, N. Christophersen and J.M. Seip, *J. Water Res.*, 1992, **28**, 1585.
14. R.L. Malcolm, in 'Humic Substances in the Aquatic and Terrestrial Environment,' eds. H. Boren and B. Allard, Wiley, London, 1991, p.369.

15. K.C. Beck, J.H. Reuter and E.M. Perdue, *Geochim. Cosmochim. Acta*, 1974, **38**, 341.
16. R.L. Malcolm and P. MacCarthy, in 'Advances in Soil Organic Matter Research. The Impact on Agriculture and the Environment,' Special Publication No. 90, ed. W.S. Wilson, The Royal Society of Chemistry, Cambridge, 1991, Ch. 2, p.23.
17. D.W. Menzel, *Deep-Sea Res.*, 1964, **11**, 757.
18. P.M. Williams, *Deep-Sea Res.*, 1967, **14**, 791.
19. Y. Miyake, K. Saruhashi, T. Kanazawa and T. Sagi, *Bull. Soc. Seawater Sci.*, 1985, **38**, 353.
20. Y. Sugimura and Y. Suzuki, *Mar. Chem.*, 1988, **24**, 105.
21. G.R. Aiken, *Environ. Sci. Technol.*, 1992. **26**, 2435.
22. D.W. Menzel and R.F. Vicaro, *Limnol. Oceanogr.*, 1964, **9**, 138.

3

Isolation, Fractionation, Functionalities, and Concepts of Structures of Soil Organic Macromolecules

By C.E. Clapp, M.H.B. Hayes,[1] and R.S. Swift[2]

DEPARTMENT OF SOIL SCIENCE, THE UNIVERSITY OF MINNESOTA, ST. PAUL, MINNESOTA 55108, USA
[1]SCHOOL OF CHEMISTRY, THE UNIVERSITY OF BIRMINGHAM, EDGBASTON, BIRMINGHAM B15 2TT, UK
[2]CSIRO DIVISION OF SOILS, GLEN OSMOND, ADELAIDE, SOUTH AUSTRALIA 5064

1 Introduction

Humic substances and polysaccharides are the major organic macromolecular components of soils, and are found in all soils which support plants and microorganisms. Both are components of soil organic matter, a term frequently used to refer to the heterogeneous mixture of soil organic materials which arise from transformations by microorganisms and chemical processes of the organic debris from soil flora and fauna. Such transformations, known as humification processes, give rise to humus substances. These bear no morphological resemblances to the plant and animal debris which provide the source materials, and have as major components the humic substances and soil polysaccharides.[1]

Humic substances can provide 80–90% of the components of humus. In the classical definitions, humic substances are composed of humic acids, fulvic acids, and humin materials. Soil humic acids are operationally defined as the components of humic substances which are precipitated at pH 1 from solution in aqueous solvents, fulvic acids are soluble in aqueous media at all pH values, and the humin materials are insoluble in aqueous solutions, regardless of the pH. However, in view of the progress which has been made in the isolation and fractionation of humic substances from soil, these classical definitions will need to be revised (see Sections 2 and 6).

There are few reliable estimates of the polysaccharide contents of humus, and the contents proposed vary widely, depending on the soil and the method of extraction. Parsons and Tinsley,[2] for example, isolated polysaccharide materials from soils amounting to 3.5–11% of the total soil organic matter, and some estimates[3] suggest that the carbohydrate content in soil organic matter can be as much as 15–20%.

There is abundant evidence[4] to show that the components of soil humus are important for:

(i)   the formation and the stabilization of good soil aggregate structures;

(ii)   the retention of plant nutrients by cation-exchange processes;

(iii)   the mobilization and the transport of metals in the soil profile, and to plant roots;

(iv)   the improvement of water entry and retention by soils;

(v)   the enhancement of the buffering capacities of soils;

(vi)   (especially for humic substances) the release of nitrogen, phosphorus, sulfur, and trace elements during the course of mineralization;

(vii)   (for humic substances) the mobilization and transport to groundwaters of anthropogenic organic chemicals;

(viii) (for humic substances) the raising of soil temperature because of increased absorption of solar radiation;

(ix)   (for humic substances) the sorption, and thereby the inactivation of some anthropogenic chemicals added to soil to control weeds, and the pests and diseases of plants;

(x)   (for humic substances) the stimulation of the growth of plants.[5]

An understanding of the structures of the colloidal components of soils, whether these colloids are inorganic or organic, is important for an awareness of the multifarious reactions and interactions which involve the colloids in the soil environment. Developments in X-ray diffraction and in electron microscopy technology during the past 60 years or so were prerequisites for the data that have led to our understanding of the compositions and structures of the clays and the oxides/(hydr)oxides of soils (see Chapters 1–9 in DeBoodt *et al.*;[6] Dixon and Weed[7]). Advances in our awareness of compositional information with regard to the organic colloids have been slower because the structures are more irregular and more complex. New and relevant types of instrumentation and approaches have been applied almost as soon as they became available to studies of these colloids. Thus, during the 1950s and 1960s applications of separation procedures, and of ultracentrifugation and electrophoresis, gave good indications of size and charge heterogeneities, and the development since then of sophisticated instrumentation and techniques such as modern gas liquid chromatography-mass spectrometry (GLCMS), pyrolysis mass spectrometry (PyMS), and various spectroscopy techniques and study procedures, such as cross-polarization magic angle spinning ^{13}C nuclear magnetic resonance (CPMAS ^{13}C NMR), Fourier transform infrared (FTIR), and electron spin resonance (ESR), as well as sensitive titration apparatus and procedures have helped our awareness of composition and structure in more recent times.

The activities of the International Humic Substances Society (IHSS), founded in Denver, Colorado, in September, 1981, has done much to advance our awareness of the compositions, structures, and functions of humic substances. Prior to that time scientists with interests in the humic substances of coal, soil, and water tended to relate only with colleagues within their own subject matter areas. At the first International Meeting of IHSS in 1983, when scientists from these different disciplines convened in Estes Park, Colorado, there was initiated a sharing of approaches, techniques, and concepts relevant to studies of humic substances that has been maintained since then.

This paper outlines the principles involved in the isolation and fractionation of the humic and polysaccharide components of soil organic matter. It summarizes the evidence which has led to our awareness of the compositional and structural features of these macromolecules and polymers, and it outlines how this awareness has improved our understanding of the reactions and interactions of these substances in the soil environment.

2 Isolation and Fractionation of Soil Organic Macromolecules

Considerations of the procedures used to isolate soil organic macromolecules should take account of the compositions of the macromolecules, and of the ways in which these are associated with each other and with the soil inorganic colloids. Because there is no evidence for genetic or biological control of the synthesis of humic substances, it would be pointless to try to isolate humic macromolecules which are homogeneous with respect to composition and structure. Therefore, the most pragmatic approaches aim to separate humic isolates from co-extracted non-humic materials, and then to fractionate on the basis of size and charge density differences.

The approach is different for soil polysaccharides. These arise from microbial synthesis processes, and from alterations to the polysaccharides and hemicelluloses of plants. Hence there is biological control of their synthesis. However, because of the vast numbers of potential sources, it is evident that the polysaccharides of soils consist of highly heterogeneous mixtures. In theory it is possible to isolate pure polysaccharides from the mixtures, but the task is daunting, and as yet no polysaccharide has been isolated from soil that satisfies all of the criteria for homogeneity.

Extraction of Humic Substances from Soil

Humic substances in the soil environment are polydisperse polyelectrolytes and the charges are pH dependent. The predominant sources of charge are in the carboxyl groups, with lesser amounts in the phenol and enol functionalities. Under acid conditions, these functional groups can be considered to behave like neutral polar molecules. However, under such conditions, cationizable groups, such as amines, become positively charged, and so humic substances can be considered to be species of variable charge.

Under the pH conditions of normal agricultural soils, humic substances are negatively charged. These negative charges are distributed throughout the macromolecular matrix, and are balanced by cations. Cations such as Li^+, Na^+, and K^+ are dissociated from the polyanion structures, and repulsion by the negative charges on the matrix gives rise to expansion of the macromolecules, and the anions and other polar groups become solvated in aqueous media. The H^+ ions are strongly associated with their conjugated bases (carboxylate, phenate, and enolate anions) to give relatively undissociated acidic groups which can give rise to inter- and to intra-molecular hydrogen bonding. In this state the shrunken molecules exclude water from the macromolecular matrix and the less polar and less charged component molecules are not solvated.

Hayes[8] has given an in-depth review of the literature and reasoning relevant to the extraction of humic substances from soil. Four criteria for effective solvents for humic substances were aptly proposed by Whitehead and Tinsley.[9] The solvents should have:

(i)   a high polarity and a high dielectric or permittivity constant to assist the dispersion of the charged macromolecules;
(ii)   a small molecular size to penetrate into the humic structures;
(iii) the ability to disrupt the existing hydrogen bonds, and to provide alternative groups to form humic–solvent hydrogen bonds;
(iv)   the ability to immobilize inorganic cations.

More recently, Stevenson[10] outlined the following criteria for the ideal extraction method. The method should:

(i)   lead to the isolation of unaltered material;
(ii)   give a humic product free from inorganic contaminants, such as clays and polyvalent cations;
(iii) give a complete extraction insuring representative fractions from the entire molecular weight range;
(iv)   be universally applicable to all soils.

Hayes[8] considered the properties of solvents which are effective for the isolation of humic substances and his data and conclusions were strongly supportive of criteria 1 and 3 of Whitehead and Tinsley. These data show that the best organic solvents for humic substances have electrostatic factor (EF) values (*i.e.* the product of the relative permittivity and dipole moment) greater than 140, and base parameter (pK_{HB}, a measure of the strength of the solvent as an acceptor of hydrogen bonds[11]) values greater than two. Thus methyl cyanide, which has an EF of 144 is a poor solvent for H^+-exchanged humic substances because it has a pK_{HB} value of 1.05, and is incapable of breaking the intra- and inter-molecular hydrogen bonds within and between the macromolecular strands.

When the classical solubility parameter (δ) of Hildebrand[12,13] is sub-divided into dispersion forces (δ_d), polar (δ_p), and hydrogen bonding (δ_h) components, further information can be obtained about the properties of solvents.[14,15] The hydrogen-bonding parameter can be subdivided into the acid or proton donor (δ_a) and the base, or proton acceptor (δ_b) parameters. The best solvents for H^+-exchanged humic acids have δ_p, δ_h, and δ_b values of the order of, or greater than six, five, and five, respectively.[8] Despite the fact that water satisfies all of these criteria, it is a poor solvent for H^+-exchanged humic acids because solution is greatest when the products of δ_a (solvent) and δ_b (solute), or *vice versa*, are maximum. The values of δ_h, δ_a, and δ_b for water are very large, indicating that the molecules are highly self associated through hydrogen bonding. No data are available for δ_a and δ_b values for H^+-exchanged humic acids, but it is obvious that these are not sufficiently great to disrupt the attractive forces between water molecules.

The δ_h value for formic (methanoic) acid is, in numerical terms, about one half that for water, but the self association of the molecules is significantly less than for water. Thus, as shown by Sinclair and Tinsley,[16] anhydrous formic acid is a good solvent for humic acids.

Some of the dipolar aprotic solvents are good organic solvents for humic substances. Although N,N-dimethylformamide (DMF) and dimethylsulf-oxide (DMSO) both have the EF, pK_{HB}, δ_h, and δ_b values to be good solvents for H^+-exchanged humic substances, based on the criteria cited by Hayes,[8] DMSO is the better of these two solvents. It is, however, necessary to have acid in DMSO because humic substances are in the anionic form (in agricultural soils, at least) and the negative charges are neutralized (predominantly) by divalent and polyvalent cations. DMSO is a good solvent for cations, but not for anions. Thus the polyanionic humic macromolecules would not be efficiently solubilized until H^+-exchanged as the result of the presence of acid in the medium. Hayes[8] has presented a scheme to indicate how solvation of humic substances in DMSO could result from hydrogen bonding between the DMSO and the carboxyl and phenolic groups in the macromolecules. Because DMSO has a non-polar as well as a polar face, it can associate with and help to dissolve the less polar moieties of humic substances.

Bremner and Lees[17] showed that up to 30% of soil organic matter could be isolated as humic substances using the sodium and potassium salts of inorganic and of organic acids. A 0.1 M sodium pyrophosphate solution, neutralized with phosphoric acid, was the best of the neutral salts, and this was followed in order of decreasing efficiencies by solutions of NaF, $(NaPO_3)_6$, Na_3PO_4, $Na_2B_4O_7$, NaCl, NaBr, and NaI. Oxalate was the most efficient of the organic salts, followed by citrate, tartarate, malate, salicyl-ate, benzoate, succinate, 4-hydroxybenzenecarboxylate, and ethanoate (see Hayes and Swift[1]).

The most efficient of the salt extractants form complexes with the polyvalent metals which insolubilize the humic substances by forming inter-

and intra-molecular bridges between the charges, or form bridges between the negative charges on the humic macromolecules and the negatively charged inorganic colloids. Such bridging effects provide important mechanisms for the retention of humic substances in soils. When the insolubilizing cations are removed the anions solvate in water. The most highly charged macromolecules will dissolve most readily, and the higher oxygen and lower carbon contents of humic substances isolated in salt solutions is proof of this.

Achard[18] used aqueous solutions of potassium hydroxide to extract humic substances from peat, and solutions of sodium or potassium hydroxide have been the solvents most widely used since then. These reagents are most effective when the humic substances are H^+-exchanged. The acid groups become ionized under the alkaline conditions and the conjugate bases solvate in the aqueous media. In principle, any system which displaces or removes the divalent and polyvalent cations, allowing the negative charges on the macromolecular matrix to repel each other, and to give rise to expansion of the molecules, should promote solvation. Aqueous diaminoethane (EDA, 2.5 M) is a good solvent for H^+-exchanged humic substances because the medium is alkaline (pH 12.6), and the acid groups dissociate to allow solvation in water.[19] In the absence of water, EDA is a poor solvent for humic substances indicating that the solute–solvent interactions are not sufficiently energetic to break the inter- and intra-molecular hydrogen bonds in the macromolecules. An alternative explanation would suggest that, on drying, the less polar groups become orientated to the exteriors of the macromolecules. These groups are not solvated by EDA, and the solvent is unable to penetrate to the internal polar structures (see also Section 2.5). However, even if EDA proved to be a good solvent it would be inadvisable to employ it for the extraction of humic substances because the nitrogen and free radical contents of the isolates are raised. Hayes *et al.*[19] and Hayes and Swift[1] have discussed mechanisms by which such N-enhancement can take place.

The combination of aqueous solutions of sodium pyrophosphate (0.1 M) and sodium hydroxide (0.1 M) dispenses with the need to H^+-exchange the humic substances, as was shown initially by Alexandrova.[20] The logic of this combination is straightforward. The pyrophosphate will complex the heavy metals, and the conjugate bases of the acidic groups (weak and strong) will solvate in water. The combination is better than the pyrophosphate solution alone (pH 10.6) because the weak acids will become dissociated at the higher pH. However, as Bremner[21] has shown, alkaline solutions enhance the oxidation of humic substances. Thus uptake of oxygen is significantly greater in pyrophosphate solution at pH 10.6 than it is when the pH is adjusted to 7.0 using phosphoric acid, and the oxidation is much greater still in solutions of sodium hydroxide. Formation of oxidative artefacts can be decreased by carrying out the extractions in an atmosphere of dinitrogen gas. The standard and reference humic and fulvic acids in the collection held by the International Humic Substances Society were isolated using alkaline (NaOH) solutions and an atmosphere of dinitrogen gas.

Extraction of Polysaccharides from Soil

Mehta *et al.*,[22] Swincer *et al.*,[23] Greenland and Oades,[3] Hayes and Swift,[1] Cheshire,[24] Stevenson,[10] and Cheshire and Hayes[25] have reviewed procedures for the isolation of polysaccharides from soil.

A useful procedure for determining the efficiency of extraction of polysaccharides, or at least of sugars or sugar-containing materials, involves determination of the sugar content in a soil hydrolysate and comparing the content of saccharide in an extract with the amount in the soil hydrolysate.

Sodium hydroxide is probably the best of the aqueous solvents for the extraction of polysaccharides from soil. Its efficiency might well be related to the varying extents of charge, arising predominantly from uronic acids, which are characteristic of many soil polysaccharides, and possibly from sulfonated polysaccharides. However, as seen above, it is also a good solvent for soil humic substances. The co-extracted humic acids are precipitated at pH 1, but there is a distinct possibility that some polysaccharides will be loosely associated with the humic acids and co-precipitated. In any event, it has been shown by Hausler[26] that saccharide materials which are not covalently linked to humic acids are present in the humic acid fraction isolated by conventional methods. When the humic acid fraction was dissolved in DMSO–HCl, and the mixture was diluted with water and passed into XAD-8 [poly(methyl methacrylate)] resin, the humic acids were sorbed by the resin and the DMSO and the polar carbohydrates and peptides were washed through.

The bulk of the polysaccharides are, however, contained in the fulvic acid fraction, and traditionally these have been considered to be grouped with the fulvic acid materials. Swincer *et al.*[23] showed that when coloured polysaccharide-containing mixtures are passed through Polyclar-AT [a poly(vinyl pyrrolidone) resin], the coloured (humic) substances are retained by the resin and the more polar polysaccharides pass through and can be recovered from the eluate. The same result is achieved by using XAD-8, and XAD-8 is the preferred resin for separating the fulvic acids from the polysaccharides and other polar substances contained in the classical fulvic acid fraction (see below).

Barker *et al.*[27] used dilute sulfuric acid to isolate polysaccharides from a sapric histosol. Dilute mineral acids are used to H^+-exchange the soil organic components, and it is clear that some of the polysaccharides are solvated in these. In the experience of Barker *et al.* dilute sulfuric acid was a better solvent for the polysaccharides than dilute hydrochloric acid, although this was opposite to that found for a mineral soil by Swincer *et al.*[23] The latter group obtained maximum yields by extracting with HCl or HF (1 M), but not H_2SO_4, followed by extraction with sodium hydroxide solution, and then, after acetylation (using acetic anhydride and concentrated H_2SO_4 at 60 °C for two hours) with chloroform. Yields ranged from 57 to 74% of the sugar contents of the soils. More recently, Cheshire *et al.*[28] have shown that methylation of the soil can lead to enhancement of the extraction of soil carbohydrate.

Dimethylsulfoxide is a good solvent for carbohydrates, but its application to the isolation of carbohydrates from soil has not been studied systematically. The uses of acidified DMSO for the isolation of humic substances from soils have been discussed already, and it is clear that polysaccharides are co-extracted. As pointed out already, the humic components are sorbed by XAD-8 resin, and the DMSO and polysaccharides pass through. However, recovery of the polysaccharides from DMSO presents problems.

'Purification' and Fractionation of Soil Humic Substances

In the context of humic substances, the term 'purification' refers to processes for the separation of these substances from the other components of soil organic matter (and especially the polysaccharides) with which they are associated in the soil environment and in the solutions used in the extraction processes. The inverted commas are used to indicate that the term does not refer to procedures which, in the chemical sense, give rise to substances that are homogeneous and satisfy the criteria for chemical purity. Fractionation refers to a subdivision of humic substances based on some property of their molecular composition.[29]

Adaptations of XAD-8 resin technology, such as that used by Hausler[26] for the removal from humic substances of non-covalently linked saccharide and peptide materials (see above), and as employed by water scientists for the removal of humic substances from waters (see Aiken[30]) can be expected to provide a useful 'purification' of soil humic substances. Considerations of the principles involved suggest that it would be appropriate to dilute soil extracts such that the concentrations of dissolved organic carbon are less than 100 mg l^{-1} (or less, to ensure that the humic acids will remain in solution when the pH is lowered to 2), and to neutralize alkaline solutions. Filtration of these solutions, using membranes which do not sorb humic substances, and with pore sizes of 0.45 microns (or better, 0.2 microns) will remove particulate materials (including suspended clays). During the course of the filtration the pores become partially blocked with suspended substances, and thus the effective pore sizes are decreased. In this way the efficiencies of the membranes are enhanced for the removal of finely divided suspended materials. The pH of the filtrate is lowered to 2 (in order to suppress ionization of the acidic groups) and, provided the solutions are adequately diluted, the humic acids can be expected not to flocculate. This acidified solution can then be pumped into XAD-8 and XAD-4 columns in series, and the humic and fulvic acid components will sorb to the first resin. There is reason to suggest that the peptide and polysaccharide components in the eluates from XAD-8 may not be sorbed by the highly non-polar XAD-4 resin because the so-called hydrophilic macromolecular acids (HMA) isolated from waters passed through these resins are low in such components. Back elution of the columns, using solvents with pH values greater than 2, allows the acid groups to dissociate, and when sufficient adjacent groups (or groups in close proximities) are ionized, the macromolecules are desorbed from the resin. Humic acids are precipitated when the pH values of the concentrated fractions of the back eluates are adjusted to 1. These are separated from the fulvic acids by centrifugation, dissolved in base, diluted, re-sorbed on XAD-8 in a desalting process, back eluted in base, then passed through an ion-exchange resin (in the H^+-form), and freeze dried. The fulvic acids and the HMA fraction are processed and desalted in the same way after the supernatant from precipitation of the humic acids, and the acidified (to pH 2) back eluates from XAD-4 are reintroduced to their respective resins.

Consideration might also be given to the uses of macroreticular resins, such as XAD-8, for the fractionation of humic substances. As mentioned, when the pH is raised, desorption will take place when the charge density of the sorbate is sufficient to give rise to repulsion between negatively charged groups on the macromolecules. A succession of solutions with increasing pH values may be used in the batchwise mode to give fractions that are broadly similar with respect to charge, or a pH gradient elution might be used to give a continuum of charged molecules. (In order to obtain molecules which are reasonably homogeneous with respect to charge density, it would be appropriate to use humic substances that are relatively homogeneous with respect to molecular sizes.) Fractions which are desorbed from the resins at the higher pH values only (pH 9 and above) would be expected to be enriched in the weaker acidic groups, such as phenols and enols.

Conventional fractionation procedures take advantage of molecular size and solubility differences. Examples of solubility differences include sequential extractions, using increasingly powerful extractants. Such may be accomplished by altering the pH of the extracting solutions,[31] or by using a sequence of solvents.[19] Another form, salting out, involves the suppression of the charge repulsion between polyelectrolytes by addition of salt. This allows the macromolecules to approach each other so that intermolecular attractive, rather than repulsive forces predominate.[32] Mention was made of the separation of humic acids from fulvic acids through precipitation of the former at pH 1. Fractional precipitation can also be achieved by isolating the fractions precipitated between pH 3 and 1. Metal ion precipitation can use a selection of divalent and trivalent metals. Use can also be made of fractional precipitation using organic solvents.[29]

Fractionation based on molecular size differences uses techniques such as gel permeation chromatography, ultrafiltration using membranes of discrete pore sizes, and density gradient or zonal centrifugation techniques.[29,31]

Fisher[33] has given an appropriate review of the principles involved in applications of gel permeation chromatography techniques. Gels from poly(acrylamide), such as the Bio Rad series, and from dextrans, such as the Sephadex series, are widely used in studies of humic substances. Progress through the gel column is retarded for molecules which are sufficiently small to enter the gel pores, and the smallest molecules are retarded most. Large molecules which do not enter the pores are eluted in the column void volume. However, humic substances can be adsorbed to varying extents by different gels, and so passage of the macromolecules through the gels may be influenced more by the adsorption processes than by entry into or exclusion from the pores. Deviation from idealized behaviour can also arise from interactions between residual negative charges on some gels and the charges on the macromolecules. Thus, repulsion between the charged species can cause low molecular weight humic substances to be excluded from the pores and to emerge in the column void volume. Use of salt in the solvent suppresses the effects of the charges and allows more meaningful results to be obtained. Typical elution patterns are given by Dubach *et al.*,[34] Swift and Posner,[35] Hayes and Swift,[1] Swift,[29] and De Nobili *et al.*[36]

Various studies have confirmed the polydisperse nature of humic macromolecules using gel permeation chromatography. The hypothetical elution pattern shown in Figure 1 is of the type obtained in the chromatography of humic substances. In order to obtain fractions that are relatively homogeneous with respect to molecular size, it is necessary to recover fractions such as those represented by I to VII in Figure 1, and to repeat the chromatography process separately for each fraction until the different fractions are eluted within their volume boundaries (indicated by the vertical lines in Figure 1). When this procedure is followed, the various fractions, with the exception of Fraction I, will be relatively homogeneous with respect to molecular size and/or shape. Fraction I, eluted in the void volume, however, will still be part of a polydisperse system, and will contain a

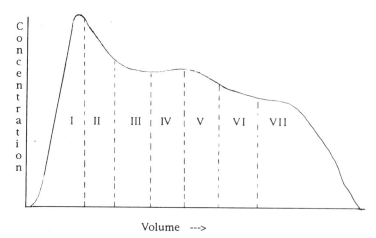

Volume --->

Figure 1 *Schematic representation of the type of elution pattern achieved when humic substances are fractionated using gel chromatography*

range of molecular sizes greater than the maximum size that enters the gel pores. Cameron *et al.*[37] utilized gel chromatography among the fractionation procedures which gave a series of components with molecular weights ranging from 2.6×10^3 to 1.36×10^6 (see Section 5).

A combination of sodium dodecylsulfate (SDS), an ionic detergent used in the fractionation of proteins, and Sephadex G-100 and Sephadex G-50 was used by Orlov and Milanovsky[38] to fractionate humic and fulvic acids. Comparison of the elution patterns from the gels in the presence and in the absence of SDS indicates that the detergent breaks the non-covalent bonding between the humic macromolecules. Fractionation of a humic acid sample, using Sephadex G-100, tris[2-amino-2(hydroxymethyl)-propane-1,3-diol]–HCl buffer (0.05 M, pH 9.0), and 1% SDS, gave three fractions having molecular weight values of the order of 150,000, 40,000, and 10,000 Daltons (the most abundant). The fraction of MW 10,000 was not observed in the absence of SDS. Good fractionation was observed for fulvic acid, using Sephadex G-50 and SDS.

Membranes of discrete pore sizes are used for fractionation by ultrafiltration. Normally the technique involves application of pressure (by means of an inert gas) to a stirred humic solution. When a range of membranes of different pore sizes are used, fractionation can be carried out in order of ascending or of descending molecular weight values. The technique does not give valid separations based on molecular size differences when there are charge interactions (attraction or repulsion) between the solute and the membrane. Stirring is important to minimize clogging of the membrane, as well as concentration polarization.

Useful separations based on molecular size differences can be achieved using density gradient and zonal centrifugation techniques.[39] Suppression of intermolecular charge repulsion by addition of electrolyte is essential in such separation processes.

Classical fractionations based on charge-density differences have used moving boundary electrophoresis, glass or cellulose column-, or paper-curtain electrophoresis,[40] polyacrylamide gel electrophoresis, isoelectric focusing, isotachophoresis, anion-exchange resins, and anion-exchange cellulose and gels.[1,4] As yet there has not been any separation in quantity of humic substances based on charge-density properties. Such separation is possible in principle, and the batchwise and gradient elution procedures using XAD resins, as referred to above, may prove to be useful for preparative work. However, whatever the procedure used, separations based on charge-density differences will require considerable commitments on the parts of those who tackle the assignment.

Fractionation of Polysaccharides from Soil

The principles described for the fractionation of soil humic substances should apply also for the fractionation of soil polysaccharides, and these have been reviewed by Hayes and Swift,[1] by Cheshire,[24] and by Cheshire and Hayes.[25] Techniques which find most widespread applications in this area include gel chromatography, for fractionation based on size differences, and anion-exchange chromatography for separations based on charge-density differences. Sephadex (a dextran) and poly(acrylamide) gels are most widely used. However, since many of the components of soil polysaccharides are charged under the conditions used, it is necessary to use dilute salt solutions to suppress the repulsion between the charges on the polysaccharides and the residual charges on the gels.

Buffered salt solutions are used in gradient elutions in ion-exchange chromatography experiments, and the ion exchangers often include diethyl-aminoethyl (DEAE) cellulose or Sephadex. (In theory, separations based on molecular size and charge-density differences are possible when DEAE Sephadex is used.) Cheshire and Hayes[25] have reproduced fractionation patterns obtained by Barker *et al.*[27] when polysaccharides isolated from a sapric histosol soil were fractionated on the basis of molecular size using a Sephadex gel, and then one of the fractions, which still had a degree of polydispersity, was fractionated on the basis of charge-density differences using DEAE-A50 Sephadex and a 0–2 M NaCl gradient for elution. Four fractions, which were clearly separated on the basis of charge-density differences, were isolated.

Use can also be made of the complexes which sugars form with boric acid to achieve fractionation of polysaccharides. Complexation, and the conferring to the complex of a borate charge, depends on the *cis–trans* relationships of hydroxyl groups on the sugars composing the polymers. Finch *et al.*[41] formed borate complexes of two polysaccharide components from the sapric histosol soil. These components were neutral on the basis of the principles used for their fractionation by gel filtration and anion-exchange chromatography. One of the polysaccharides, which had a molecular weight of *ca.* 50,000 Daltons (determined by ultracentrifugation measurements) moved as a single component in moving boundary electrophoresis when phosphate (pH 7.0), barbiturate (pH 9.0), and borate (pH 9.1) buffers were used. The other moved as a single component in the phosphate and barbiturate buffers. It was, however, split into four negatively charged fractions in borate buffer, and that was clear evidence for inhomogeneity. Clapp[42,43] made use of this principle to isolate a neutral polysaccharide using the Kirkwood cell and paper and column electrophoresis, and he used column electrophoresis and borate buffer to separate the component sugars in the hydrolysate of the polysaccharide.

3 Functionality in Soil Organic Macromolecules

The physico-chemical properties of the organic macromolecules of soils depend on their sizes and shapes, and on the functional groups of the component molecules. The polar and the non-polar functionalities are important for solvation, for ion exchange, and for complexation properties, and the non-polar functionalities are relevant to the hydrophobic properties, and to the binding of non-polar and sparingly soluble organic chemicals to the soil organic constituents.

Functionality of Soil Humic Substances

A combination of wet chemistry analytical procedures, data from spectroscopy and from titration analyses, and identifications of products of degradation processes (see Section 4) can give useful indications of the functional groups contained in humic substances.

Wet Chemistry Methods. Hayes and Swift[1] have reviewed the strengths and weaknesses of the traditional wet chemistry methods used for the detection of some functional groups in humic substances. The uses of diborane (B_2H_6), which liberates gaseous hydrogen even from sterically hindered acidic groups in H^+-exchanged humic preparations, and of titration, where excess of base is back titrated after humic samples are treated with barium hydroxide, have been widely used in determinations of total acidity. The barium hydroxide method can, of course, give high values because of the likelihood that some barium is held by chelation or complexation processes. Carboxyl groups, the major contributors to the acidities of soil humic substances, may be determined by decarboxylation, using the $CuSO_4$– quinoline reagent (Hubacher[44]), and the calcium acetate (ethanoate) method (Schnitzer and Gupta[45]) in which calcium salts of the acids are formed and the ethanoic acid liberated is measured by titration. For this procedure to be effective it is important that all of the Ca^{2+} ions should have access to all of the carboxyl groups, and that exchange should not take place with hydrogen ions of the phenolic groups. However, as Schnitzer and Gupta have shown, calcium ethanoate reacts with the more strongly acidic phenolic groups, and so this method can be expected to give over estimates of the carboxyl group contents. In addition, the 'cross-linking effect' of Ca^{2+}, where the divalent cation neutralizes carboxyl groups from different macromolecular strands, or even from different regions in the same strand, restricts access to non-exchanged sites in the macromolecular matrix. Such structural changes could cause low values to be obtained for the carboxyl group content. The phenolic acidity is often considered to be the difference between the total acidity and the carboxyl group content.

Hayes and Swift[1] have reviewed procedures which determine carbonyl and quinone groups by measuring increases in N contents of derivatives formed when the humic substances are treated with hydroxylamine, with phenylhydrazine, and with 2,4-dinitrophenylhydrazine. Other approaches measure the amounts of unreacted reagents. For example, Schnitzer and Skinner[46] formed the 2,4-dinitrophenylhydrazone derivatives of humic substances by refluxing the humic samples with an excess of the 2,4-dinitrophenylhydrazine reagent, and the carbonyl content was estimated from the amount of unreacted reagent determined from the data for the polaro-graphic reduction of the nitro groups. There is good reason to consider that quinones, arising from oxidations of phenols, are important function-alities in soil humic substances, and Flaig's Group (at Braunschweig, Germany; see Flaig et al.[47]), attached considerable significance to hydroxy-benzene and to quinone components in humic structures. (Quinones in humic substances may well have significance in the binding of anthro-pogenic organic chemicals which can form charge-transfer complexes.) There is, however, no unambiguous method available for determinations of quinone groups in humic macromolecules, because the types of derivatives which form with the carbonyl groups of quinones will also form with the other carbonyl functionalities (such as those in the more usual aldehyde, ketone, and ester structures; see Hayes and Swift,[1] p. 200).

Spectroscopy Procedures. Nuclear magnetic resonance (NMR) of humic substances in the liquid (see reviews by Wershaw,[48] Steelink et al.[49]) and in the solid states (see reviews by Wilson,[50] and by Malcolm[51]), infrared (see MacCarthy and Rice[52]), and electron spin resonance (ESR; see Senesi and Steelink[53]) are the three spectroscopy procedures used most widely for studies of the functionality of soil humic substances. Other spectroscopy procedures, such as Raman spectroscopy, ultraviolet-visible spectroscopy, fluorescence spectroscopy, X-ray photoelectron spectroscopy, and Möss-bauer spectroscopy have more limited uses in studies of humic sub-stances.[54] However, as Hayes et al.[55] have emphasized, humic substances represent 'the epitome of molecular heterogeneity and complexity' and 'any attempt to interpret rigorously the data that result from the applica-tion of a given technique to humic substances may be likened to attempting to identify the thousands of persons in a stadium when they all shout their names in unison'.

Derivatization of humic samples can allow more meaningful interpreta-tions of spectroscopy data. For example, Leenheer and Noyes[56] and Bloom and Leenheer[54] have described derivitization procedures that assist assign-ments in infrared and NMR spectroscopy of humic substances. Problems of spectral overlap of oxygen-containing functionalities can be partially re-solved in this way, and Leenheer and his colleagues have shown how derivitization techniques in association with infrared spectroscopy can be used to identify carboxyl, ester, ketone, and hydroxyl functional groups in humic substances from freshwater environments. It should be possible to apply the same approach to studies of the infrared spectroscopy of humic substances from soils.

There was considerable excitement when cross-polarization magic angle spinning (CPMAS) ^{13}C NMR was introduced over a decade ago to studies of functionalities of soil humic substances. Some who used the technique have been overly ambitious with their interpretations. The spectra can show the presence and the relative abundances of aliphatic and aromatic constituents and of carboxyl/ester groups, and indicate that phenols, ether-type functionalities (and specifically methoxyl groups), and carbohydrates or carbohydrate-derived materials are present, but as yet they cannot provide detailed information of how these functionalities are arranged in the macromolecular humic substances.

There are 'pitfalls' in the interpretations of NMR spectra, and some workers have been over enthusiastic in their assignments of chemical shift data. Quantitative interpretations have been placed on data from spectra that, at best, could be considered in a qualitative manner only. For applications of liquid-state ^{13}C NMR to studies of humic substances, it is essential to use adequate time delay between pulses for complete T_1 and T_2 relaxations, and account must be taken of the possible influences of solvents, the nuclear Overhauser enhancement, and the inverse gated decoupling sequence. For solid-state ^{13}C NMR spectroscopy, it is important also to have appropriate time delays for complete T_1 and T_2 relaxations, to have adequate speeds in the magic angle spinning to prevent spinning side-bands, to have sufficient high-power proton decoupling, and to have the proper contact time to maximize the cross polarization. The presence in the sample of paramagnetic ions, and especially iron and copper, influences the mode of carbon relaxation and the cross-polarization efficiency within the sample. Therefore the ash contents of samples, and especially the amounts of paramagnetic species in the ash, can strongly influence the extents to which the carbon in samples of humic substances can be observed by this technique. It is possible to observe up to 97% of the carbon when the contents of ash and of paramagnetic species are low.

Recently Malcolm,[57] in a study of aquatic humic substances from Lake Skjertvatjern, Norway, has stressed the importance of four factors for quantification of solid-state ^{13}C NMR spectra. These are:

(i) favourable T_{CP} and T_{1p} time constants in which, in variable contact time (CT) experiments, T_{CP} is the cross-polarization time magnetism to be transferred from H spins to C spins, and T_{1p} is the proton relaxation time;

(ii) a favourable repeat time in the ^{13}C NMR experiment. (The repeat time should be at least three to five times longer than the longest T_{1p} in the molecule);

(iii) the percentage of carbon in the spectra;

(iv) the determination of organic free radicals and/or inorganic paramagnetic species in the sample which may cause losses of the ^{13}C NMR signal in selective parts, or in all parts of the spectrum.

It is therefore important to carry out an in-depth study of the ^{13}C NMR relaxation parameters in order to quantify the spectral data. Such studies have been rare. Thus many of the so-called quantitative data in the literature are, at best, only qualitative.

From the plots of contact time *versus* the relative peak heights for chemical shift data ranging from 38 to 172 ppm, Malcolm was able to show that, in general, the signal gradually built to a maximum for a contact time of 0.5–0.8 ms, held a plateau of values in the range of 0.7–1.0 ms, and then diminished for contact times in excess of 1 ms. The T_{CP} values for both the humic and fulvic acids were relatively low, ranging from 0.15 to 0.53 ms for the fulvic acids, and from 0.14 to 0.37 ms for the humic acid samples. The cross-polarization times were, as expected, fastest for the aliphatic (hydrogen-rich) portions of both the humic and fulvic acid macromolecules. Under the experimental conditions used, contact times in the range of 0.9–1.1 ms enabled satisfactory quantitative data to be obtained.

CPMAS ^{13}C NMR spectra are shown in Figure 2 for humic acids isolated from the silt and clay fractions of a Mollisol soil to which maize residues were added (R), or were not added, but were removed with the crop during a period of about 18 years (NR). The spectra are essentially the same for the two humic acid samples (from the R and NR treatments) from the clays. It is evident also that there are similarities between the humic acids from the silts and clays, but the relative abundances of the components represented by the different chemical shift data are different.

The chemical shift data in the 25–35 ppm range is indicative of aliphatic carbon, and would suggest the presence of methylene groups in alkyl chains. The shifts in the 50–55 ppm region are likely to be attributable to methoxy and ethoxy groups, to other ethers, and to ester-type groups. The distinct shift in the 65–70 ppm range could be attributable to aliphatic carbon moieties linked by oxygen (ether linkages), and/or to carbon bonded to secondary alcohol-type structures, including saccharides. Sugar analyses indicated that the carbohydrate contents of these humic acids were low, and that was confirmed by the small shoulder in the chemical shift region at about 100 ppm (which might be attributable to saccharides or to saccharide-derived materials). Clear cut evidence for aromaticity is shown in the chemical shift region between 120 and 140 ppm. There is a distinct difference between the spectra for the humic acids from the silt and from the clay fractions in the chemical shift regions at about 150 ppm. That shift is characteristic of phenol functionalities, and these are not evident in the spectra for the clay humic acids. The chemical shift at 160–180 ppm is attributable in the main to carboxyl carbon, but other contributors to that range include carbonyl and ester carbons. The chemical shift at 230–250 ppm is likely to be attributable to carbonyl carbon.

It would be wrong to try to interpret the NMR data that are available for humic substances in terms of detailed structures. However, when spectra are obtained under well controlled conditions they provide very useful information about aliphatic to aromatic ratios, and, as indicated above, indications of the presence of certain types of functionalities. The spectra provide excellent 'finger print' techniques for comparing samples from different origins, and isolated by different procedures, and subjected to different treatments, *etc.*

Titration Procedures. Potentiometric titration procedures provide information about acidic groups in humic substances. The data show that some
strong acid groups are contained in the macromolecules, and these are
likely to arise from activating substituents close to carboxyl groups. It is
clear that most of the acidity can be attributed to carboxyl groups, but the
titration data show that weakly acidic groups are also present, and these
would include phenolic functional groups. However, NMR spectroscopy
does not always confirm the presence of phenols, as is evident for the
spectra for the humic acids from the clay fractions of the Mollisol soil in
Figure 2. The absence of evidence for phenols in the spectra appears to be
especially relevant to humic substances in well humified soils, such as some
Mollisols. That might suggest that phenols in such humic substances are
oxidized more extensively to quinones than they are in samples which show
chemical shift data characteristic of phenols. It is possible also that the

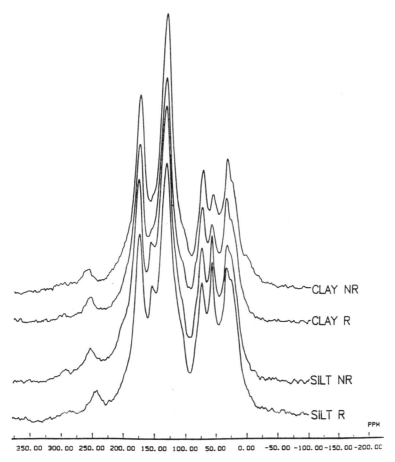

Figure 2 *CPMAS ¹³C NMR spectra of humic acids isolated from the silt and clay
fractions of a Mollisol soil to which maize residues had been added* (R) *or
not added* (NR) *over a period of about 18 years*

chemical shift for phenols is not evident in humic acids from some soils because of a masking effect which has not been addressed.

Degradation Procedures. Degradation procedures are used for studies of the component molecules of humic substances. Identifications of products in the degradation digests, and interpretations of the mechanisms of the degradation reactions can allow deductions to be made about functional groups in humic substances. This matter is discussed further in Section 4.

Functionality of Soil Polysaccharides

The component sugars of soil polysaccharides can be determined quantitatively using a variety of analytical procedures and a range of sugar derivatives of the hydrolysates of the polysaccharides.[24,25,28] The ratios of the sugars allows deductions to be made of the contributions to the functionalities of carboxyl groups (in uronic acids), and of acetylamino groups (in acetylated amino sugars).

4 Component Molecules of Soil Organic Macromolecules

By component molecules is meant the molecular entities which compose the polymer or macromolecular structures. Such molecules are readily determined for hydrolysable structures, such as the peptide bonds linking the amino acids of proteins, or the glycosidic linkages between sugars in oligosaccharides and polysaccharides. The situation is far more complex for the linkages between the component molecules of humic substances.

The Component Molecules of Soil Humic Substances

Mention was made in Section 3 of the kinds of information that spectroscopy procedures can provide about functionality in humic macromolecules. It is clear that, although such information is important, it cannot provide any accurate indications about the types of molecules which compose the macromolecules. Such molecules are often called 'building blocks' in macromolecular chemistry. The most reliable information about the composition of the macromolecules has come from degradation procedures and the identification of the compounds released in the digests when the macromolecular structures are broken down. These procedures include hydrolysis, oxidative and reductive processes, and pyrolysis.

Information from Hydrolysis Processes. Component molecules of polysaccharides, proteins, and nucleic acids are identified in digests after degradation by hydrolysis of humic substances. There is genetic or biological control in the synthesis of proteins, polysaccharides, and nucleic acids, and water is released as the linkages between the component structural units are formed. However, the processes of formation or genesis of humic substances are varied and complex, and the linkages between the components of the major structural units (sometimes referred to as the 'backbone') are difficult to cleave. For the most part, these consist of sugars and amino acids released during hydrolysis. The sugars, amino acids, and the traces of the nucleic acid bases identified in the digests could well be components of oligosaccharides or polysaccharides, and of peptides and nucleic acids co-precipitated with the humic acids, or contained in the fulvic acid fraction in instances where that fraction was not treated with a 'cleansing' resin, such as XAD-8 (see Section 2). It is evident that some sugar and amino acid containing components are covalently linked to the humic 'backbone' (see Hayes *et al.*[55,58]), but this linking process would take place after the saccharides and peptides were biosynthesized in separate processes.

Parsons[59] has reviewed the mechanisms of hydrolysis relevant to humic substances, and he has listed the types of products identified in the digests. Up to 50% of the mass of the humic acids can be lost as CO_2 and as soluble molecules when the humic acid fraction is hydrolysed under reflux conditions in 6 M HCl. The soluble molecules include the sugars, amino acids, and small amounts of purine and pyrimidine bases, as referred to above, in addition to phenolic compounds. Hydrolysis processes, however, give rise to the formation of browning reaction products as the result of condensation reactions between the sugars and amino acids released from carbohydrate and peptide structures. Although the losses of CO_2 are significant (through, for example, decarboxylation of β-keto acids), the total acidities of the macromolecules are not appreciably changed as the result of the hydrolysis processes. That would suggest that new acid groups (*e.g.* from esters) are formed during hydrolysis.

Information from Reductive Degradation Processes. Hayes and Swift[1,4] and Stevenson[60] have described procedures and mechanisms for the degradation of humic substances under reductive conditions. Degradations with sodium amalgam have provided the most useful of the information obtained from reductive procedures in so far as the composition of humic macromolecules is concerned. Figure 3 (from Hayes[61]) gives representations of the types of structures found in the digests of degradations with sodium amalgam.

There were indications from infrared spectroscopy that aliphatic components were present in some of the digests. However, the analytical procedures available at the time when the work was done did not permit identification of these, and the compounds identified were mostly phenols and their derivatives. From model studies it was shown that phenolic ether, and biphenyl structures bearing activating substituents (hydroxyl, methoxy, methyl) *ortho* and/or *para* to the carbon linking the phenyl groups were

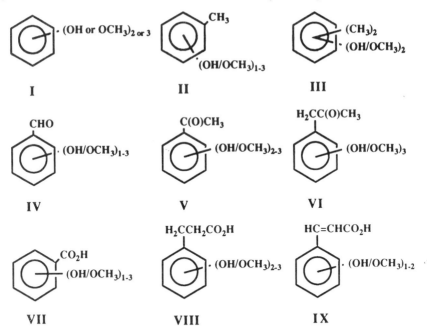

Figure 3 *Types of structures identified in the digests of degradations of humic acids with sodium amalgam preparations (from ref. 61)*

cleaved. Aliphatic to aromatic ether linkages would be broken in a similar way.

Several of the aromatic structures identified in digests of degradations with sodium amalgam had carboxyl as a substituent on the aromatic nuclei (structures of type VII), but there was not, in any instance, more than one carboxyl group on any aromatic ring. Because degradation was reductive, these acid groups were unlikely to be artefacts of the degradation process. It is probable also that some products of types I, II, and III were derived from benzene carboxylic acids bearing two to three activating substituents on the aromatic nuclei. Such substituents would promote decarboxylation under the reflux conditions used in the degradation processes.

The aldehyde and ketone functional groups in structures of types IV, V, and VI are unlikely to be artefacts of the sodium amalgam degradation process, and can be considered to give indications of the ways in which these functionalities are present in the macromolecular structures.

Compounds of types VI, VIII, and IX can be considered to be derived from phenylpropane-type structures, and the substituents (OH and OCH$_3$), when located on ring positions 3- and 4-, or 3-, 4-, and 5- would suggest origins in lignin derived materials. Some such compounds identified in the digests had substituents in the 3- and 5-ring positions only, and that would suggest that the parent structures arose from microbial synthesis processes.

Sodium in liquid ammonia provides another useful reductive procedure for the degradation of humic substances. There have been no recent studies of applications of this procedure, and the earlier studies did not investigate the degradative mechanisms involved. The results obtained would, however, suggest that ether linkages are important in humic structures.

Zinc-dust distillation of humic substances was introduced by Cheshire *et al.*[62,63] and this procedure, as well as zinc-dust fusion was subsequently used by Hansen and Schnitzer.[64,65] The major digest products were fused aromatic structures, some of which were N-heterocycles (see Hayes and Swift[1]). Yields were invariably low, often less than 1% of starting material, and the recovery (as a pale yellow oil) of 3% of starting materials was considered to be good. The yields obtained by Hansen and Schnitzer were of the order of 0.6–0.7% for humic acid, and those for fulvic acids were less. They calculated that fused aromatic structures could account for 25 and 12%, respectively, of the components of humic acids and fulvic acids, if it was assumed that recovery of products was 10%, and if the calculations were made on a 'functional group-free' basis.

Interpretations of such data led Cheshire *et al.*[63] and Haworth[66] to propose that humic macromolecules have at their 'core' a polycyclic aromatic structure to which is attached simple phenols, metals, peptides, and carbohydrate structures.

In their later communication, Cheshire *et al.*[63] showed that 3,4- and 3,5-dihydroxybenzoic acids, furfural (from the dehydration of pentose sugars), and polymers (from quinones) also gave polycyclic aromatic structures during zinc-dust distillation at 500–550 °C. At 400 °C only small amounts of anthracene were obtained from the hydroxybenzenes, and no aromatic structures were detected in the distillates from furfural and polymers of *ortho*- and *para*-benzoquinone. Because the same products were isolated in the same proportions from the digests at the higher and at the lower temperatures for the humic substances, it was concluded that the fused aromatic compounds identified in the digests were largely from the humic substances.

The evidence from zinc-dust distillation and fusion processes does not allow us to conclude with confidence whether or not the fused aromatic structures identified were components of the humic macromolecules. It would be appropriate, for example, to subject compounds released by other reductive techniques, as well as the undegraded residual materials to zinc-dust distillation and fusion reactions at 400 °C and to compare the yields of fused aromatic structures from both types of substances. Such experimentation would allow some conclusions to be reached with regard to the extent of artefact formation resulting from the degradation processes.

Information from Oxidative Degradation Processes. There have been well over 100 compounds identified in the digests of the various oxidative processes used for the degradation of humic substances. That does not necessarily indicate that there are over 100 different 'building blocks' in the macromolecules. In any particular oxidative process several different products could arise from the same precursor, and especially when the time of reaction, the temperature, the concentration of reagents, *etc.*, are altered.

Figure 4 provides examples of the types of compounds that have been identified in the digests for the degradation of humic substances with alkaline permanganate, and with alkaline copper(II) oxide. For details of the actual structures identified, see Hayes and Swift[1] and Griffith and Schnitzer.[67] Benzenedi- and benzenepoly-carboxylic acids predominated in the digests where permanganate was used, and benzenetri-, tetra-, and penta-carboxylic acids were most prominent in the digests of alkaline copper(II) oxide. Aliphatic dicarboxylic acids (structure type XIII), ranging from ethanedioic (where $n = 0$) to decanedioic (where $n = 8$) acids were also abundant components of the digests. Aliphatic tricarboxylic acids, long-chain monocarboxylic acids (structure type XI, where $n = 10–22$, and higher), as well as long-chain aliphatic hydrocarbons (structure type X, where $n = 12–38$) have also been identified, especially in the digests of alkaline copper(II) oxide degradation reactions.

The abundance of benzenecarboxylic acid structures in the permanganate digests might be interpreted in terms of origins in fused aromatic structures. However, the same acids are found in the digests of degradations with alkaline copper(II) oxide, and such reagents would not degrade fused aromatic structures to benzene polycarboxylic acids. It is therefore likely that the carboxylic acids were formed from oxidations of aliphatic side chains. There is the possibility also that the carboxyl groups could arise from carbonylation processes under the reaction conditions used.

Carboxyl groups are generated during oxidative cleavages of alkenes. Thus alkene structures would be plausible precursors for structures of type

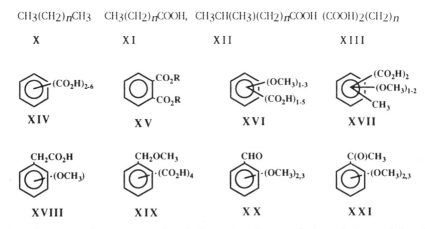

Figure 4 *Types of structures identified in the digests of degradations of humic substances with alkaline permanganate and with alkaline copper (II) oxide (from ref. 61)*

XIII identified in the digests, where n represents the number of CH$_2$ groups separating the double bond functionalities. The carboxyl groups could also arise, of course, from the oxidation of aldehyde and of primary alcohol functional groups.

It was necessary to derivatize the carboxyl and phenol hydroxyl groups for identification by gas liquid chromatography-mass spectrometry (GLCMS). It is certain that the acid groups would have been present as the sodium salts in the digests. However, since the methyl derivatives were made, it cannot be stated with certainty whether or not the methoxy substituents were present in the native aromatic structures, or were formed by methylation of phenolic hydroxyl groups. From one to three methoxyl groups were identified in some of the methylated aromatic digest products, and these were variously accompanied by one to five carboxyl groups (structure type XVI).

It is tempting to infer that the phthalate esters (structure type XV) were derived from plastics. However, these were present in digests where contamination with plastics was very unlikely, and it must be assumed that such structures were formed from components that are indigenous to humic substances.

It is clear that oxidation was incomplete for structure types XVIII, XIX, XX, and XXI because aliphatic side chains survived in the digests.

Information from Degradations with Sodium Sulfide and with Phenol.
Hayes and O'Callaghan[68] have discussed the mechanisms of reaction of sodium sulfide with humic substances at elevated temperatures. It is clear that some oxidation of the digest products can take place under the reaction conditions.

Figure 5 shows the types of products that have been identified in the digests. These include aliphatic alcohols (where $n = 1$–4 in structure type XXII), acids (especially ethanoic, XXIII, possibly from carbohydrates associated with the humic structures, and those represented by structure type XXII, where $n = 1$–4), as well as hydroxy and keto acids (structure types XXIV, XXV, and XXVI). The aromatic structures were very different from those identified in the classical oxidative degradation reactions. Digest products were methylated for identification by GLCMS, and it was seen that most of the aromatic structures had one or two methoxy substituents, one or two methyl and other aliphatic substituents, and with the exception of compound XXXI, only one carboxyl group was attached to the benzene nucleus. This contrasts with the benzenecarboxylic acid structures found in the digests where permanganate and alkaline copper(II) oxide were used.

The formation of quinone methide structural intermediates is important to the mechanisms of degradation with Na$_2$S at elevated temperatures. Hayes and O'Callaghan[68] have shown how a methyl substituent would arise from a hydroxy or ether functional group on a carbon attached to the

XI, XIII, $CH_3(CH_2)_nOH$ CH_3COOH $CH_3CH(OH)COOH$ $CH_3(O)(CH_2)_2COOH$

$$\textbf{XXII} \qquad\qquad \textbf{XXIII} \qquad\qquad \textbf{XXIV} \qquad\qquad \textbf{XXV}$$

$CH_3(O)C(CH_3)_2CH_2COOH$

$$\textbf{XXVI} \qquad\qquad\qquad \textbf{XXVII} \qquad\qquad\qquad \textbf{XXVIII}$$

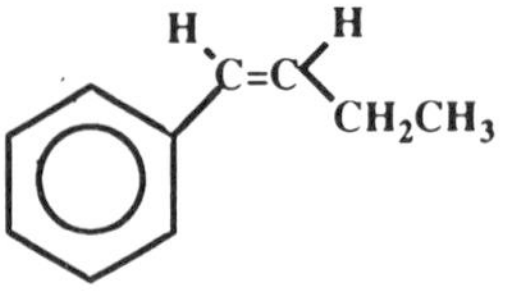

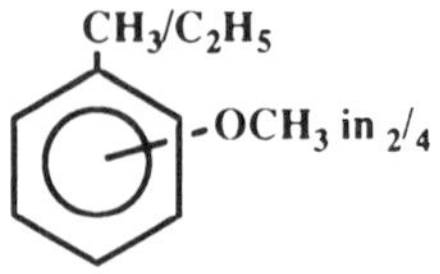

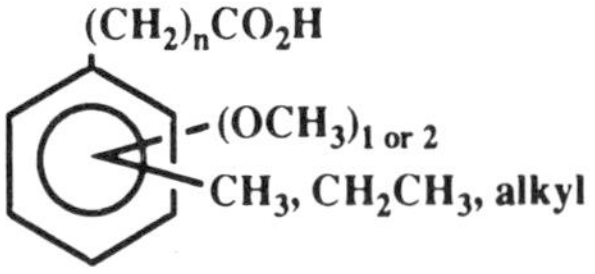

$$\textbf{XXIX} \qquad\qquad\qquad \textbf{XXX} \qquad\qquad\qquad \textbf{XXXI}$$

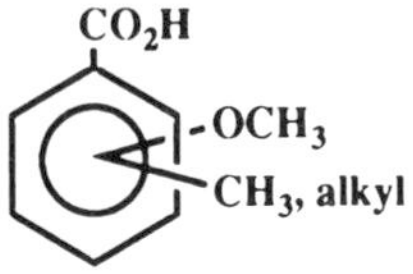

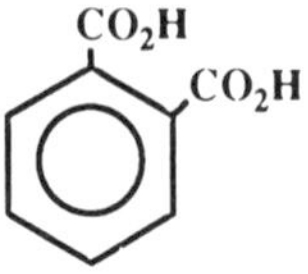

Figure 5 *Types of structures identified in the digests of degradations of humic substances with* Na_2S *(10% w/v) at 250 °C (from ref. 61)*

aromatic ring, and *ortho* or *para* to a (phenolic) hydroxyl substituent. (Methyl substituents can also form as artefacts on the ring positions during the course of the methylation of phenols.) In addition to methoxy, the alkyl substituents (other than methyl), detected in structure types XXVIII, XXIX, and XXX, could not have been artefacts formed during methylation. Ethyl, and other alkyl substituents could arise from appropriate secondary alcohol and ether structures attached to the carbon substituent on the aromatic ring, and *ortho* or *para* to a phenolic hydroxyl. The evidence provided by Hayes and O'Callaghan would suggest that the quinone methide intermediate played an important part in the mechanism of cleavage of aromatic moieties from the humic macromolecules. In their view, many of the products identified could have had origins in phenyl-propane structures of the types associated with lignins, and in which two or three hydroxy or ether substituents were present in the propyl side-chain group.

Heredy and Neuworth[69] introduced phenol, heated for 24 h at 100 °C, and in the presence of BF_3, for the depolymerization of coal. Later Ouchi et al.,[70] and Ouchi and Brooks[71] refluxed brown coal with phenol plus *p*-toluenesulfonic acid and postulated that the process released aliphatic structures linking aromatic nuclei. Hayes and O'Callaghan[68] have reviewed the relevant literature and the known mechanisms, and have described studies which show that the products of degradation are more complex than those considered previously. They concluded that there are several possible origins for the digest products of degradations of humic substances with phenol plus catalyst. It is clear that the reflux reaction conditions can give extensive degradation of humic substances, and there is a need to identify the numerous products known to be present in the digests and, on the basis of even the limited degradation mechanisms known to apply, to relate these products to possible origins in the humic macromolecules.

Information from Boron Trifluoride–Methanol Transesterification. The BF$_3$–MeOH depolymerization method, which was successfully used in structural studies of plant cutins and suberins (Kolattukudy[72]), has been used by Almendros and Sans[73,74] for studies of the composition of humic acids and humin. The procedure is mild; it provides a selective removal of the humic constituents linked by ester functional groups, and it allows the labile (derivatized) structures to be identified. Yields of identifiable digest products are of the order of those from degradations with permanganate and with alkaline copper(II) oxide (30–35%). Products of the degradations of humins[74] consist of a variety of monobasic straight-chain and branched fatty acids, long-chain dicarboxylic acids, dihydroxymonobasic acids, tri-hydroxymonobasic acids, benzenedi-, tri-, and tetra-carboxylic acids, methoxy benzenecarboxylic acids, and di-, tri-, and tetra-methoxy benzenecarboxylic acids, and a variety of miscellaneous acids. For humins, the compounds identified were typical of the depolymerization products of plant components, such as cutins and suberins.

Information from Pyrolysis. Pyrolyzates of alkali-extracted soil humic acids give compounds rich in substances of polypeptide, of lignin, or of polyphenol origins, and to a lesser extent some of the compounds would appear to have origins in polysaccharide or in the so-called 'pseudopolysaccharide' structures.[75] (The term 'pseudopolysaccharide' was introduced by pyrolysis scientists to cover possible origins in macromolecules for pyrolysis products such as furan, methylfuran, dimethylfuran, furfural, and methyl furfural structures.) In the view of Bracewell *et al*,[75] the evidence for 'pseudopolysaccharide' structures is especially strong for the fulvic acid fractions. However, the experiments referred to were carried out on samples which had not been 'cleansed' of associated peptide and carbohydrate materials, and it is likely that the evidence for structures with origins in carbohydrate-derived materials would be less strong for fulvic acids subjected to treatment with XAD-8 resin (see Section 2). This thesis is substantiated by the fact that evidence for peptide and carbohydrate-derived structures was diminished when the samples were first hydrolysed in 6 M HCl. The abundance of phenol pyrolysis products, derived from altered lignins and microbially synthesized polyphenol substances, was increased by the hydrolysis treatment.

More recently, Schulten *et al.*[76] pyrolized (at 500 °C) humic acid samples in a Fisher Curie-point pyrolyzer attached to a GCMS system with a 46–400 Dalton mass range. The major thermal products were benzene and alkylbenzenes, and C$_1$ to C$_{13}$ *n*-alkylbenzenes were especially prominent, and there was evidence also for branched members of the same series. Some tri- and tetra-methylbenzenes, alkylnaphthalenes, and alkylphenanthrenes were also detected. The information indicated that isolated aromatic rings are linked covalently to aliphatic chains. This type of linkage in humic structures has been known for a considerable time, based on studies of degradations with phenol (as referred to above). The presence of some naphthalene and phenanthrene structures is of interest, but it is possible that these were formed as artefacts, because the chemical degradation techniques (apart from the drastic zinc-dust distillation and fusion procedures) have not provided any substantial evidence for fused aromatic structures.

Schulten *et al.* concluded that the alkyl–aryl compounds detected in the pyrolyzates were released from an alkylaromatic structural network in the macromolecules. They proposed a tentative network structure which was very much at variance with the loose random coil arrangement suggested by the careful physico-chemical studies of Cameron *et al.*[37] and discussed in Section 5. Furthermore, there were no polar groups on the macromolecular structure proposed, and such is at variance with the overwhelming evidence for polar substituents in the macromolecular structures.

The Component Molecules of Soil Polysaccharides

The glycosidic linkages in polysaccharide structures are readily hydrolysable in acid. These linkages are also susceptible to enzyme hydrolysis. Thus it can be expected that most polysaccharides will have a transient existence in the soil. However, polysaccharides must have a degree of protection in soils where they are present in considerable abundance, and that protection is likely to be steric, whereby the glycosidic linkages are shielded from enzyme attack by the conformations adopted by the polysaccharides, or by associations which the polysaccharides can form with other components of soils, such as the clays, (hydr)oxides, and humic substances.

Table 1 (from Cheshire and Hayes[25]) gives the sugar contents of a whole soil, the relative abundances of each sugar in the soil hydrolysate, and in a polysaccharide isolated from an alkali extract of the soil. Glucose is present in greatest abundance in the hydrolysate of the whole soil, but the content in the polysaccharide isolate is similar to that of galactose and mannose. That would suggest that considerable amounts of the polysaccharide materials in the soil were present as cellulose-type materials, and these would not be isolated in the alkaline extract.

The high relative contents of the pentose sugars arabinose and xylose suggest plant origins for these sugars. (Glucose, xylose, and arabinose usually constitute the major sugars of plant tissues.) Microorganisms synthesize arabinose in the L- and D-configurations, whereas this sugar is synthesized only in the D-configuration by plants. Cheshire and Thomp-

Table 1 *Contents of sugars in the hydrolysate of an arable soil and the relative percentages of these sugars in the hydrolysates of the soil, and of a polysaccharide isolated from that soil (from Cheshire and Hayes[25])*

Sugar	Sugar content of whole soil (mg g^{-1})	Sugar as % of all sugars in soil hydrolysate	Sugar as % of sugars in soil polysaccharide hydrolysate
Galactose	1.4	07.3	15.3
Glucose	5.2	27.2	17.9
Mannose	1.5	07.8	15.0
Arabinose	1.5	07.8	06.2
Ribose	0.05	00.3	—
Xylose	1.6	08.3	08.8
Rhamnose	0.8	04.2	05.1
Fucose	0.3	01.6	04.7
Glucosamine	0.5	02.6	06.2
Galactosamine	0.5	03.6	08.0
Glucuronic acid	2.6	13.6	04.7
Galacturonic acid	3.0	15.7	08.0

son[77] have shown that at least 90% of arabinose in one soil studied was in the L-configuration. Galactose can also be present in considerable abundance in some plants, and L-fucose (or 6-deoxy-L-galactose) is present in lesser amounts. There usually are negligible amounts of mannose and of L-rhamnose (6-deoxy-L-mannose) in plants. The hexose sugars glucose, mannose, and galactose, the amino sugars glucosamine (from acetylglucosamine) and galactosamine (from acetylgalactosamine), and the two uronic acids, glucuronic acid and galacturonic acid, could be from animals. Chitin, from insects, would be an obvious major source of acetylglucosamine. Ribose, in small amounts, would be present in animal and plant tissues.

Experiments by Cheshire *et al.*[78,79] have shown that the label in ^{14}C-tagged substrates that are readily metabolized by microorganisms, such as glucose, xylose, and starch, is predominantly distributed among the hexoses glucose, galactose, and mannose, in the deoxyhexoses rhamnose and fucose, and, with weaker labelling, in the pentoses arabinose and xylose. These findings suggest that microorganisms are largely responsible for the hexoses and deoxyhexoses of soils. When such labelling experiments are extended for up to two years, a high level of activity develops in the deoxyhexoses, and in particular in rhamnose. However, the composition of the newly synthesized carbohydrates bears only a superficial resemblance to that of the whole soil carbohydrate. That could indicate that insufficient time was given to allow for the differential degradation rates of the synthesized carbohydrates, but it could also indicate that microbial synthesis is only a part of the origins of soil carbohydrates.

5 Sizes and Shapes of Soil Organic Macromolecules

An awareness of shapes and sizes, and of composition, is important in considerations of the reactions and interactions of the organic macromolecules that are indigenous to soil.

Molecular Sizes and Shapes of Humic Macromolecules

Hayes and Swift[1,4] and Swift[80,81] have discussed the uses of ultracentrifugation for determinations of the sizes and shapes of humic macromolecules, and much of the discussion which follows is based on these references. Considerations of the uses of ultracentrifugation must take account of the polyelectrolyte nature of the humic substances at the pH values at which the experiments are carried out. Humic macromolecules in solution will have negative charges distributed along their lengths, which repel each other. (Fulvic acids at low pH values do not necessarily conform to this concept because these substances are soluble at the pH values at which ionization is suppressed.) These charges are balanced by cations which restrict, but do not prevent molecular expansion. In addition there is intermolecular repulsion between the negatively charged molecules, and repulsion or attraction, depending on the sign of the charge, between the humic macromolecules and other charged surfaces. These charge effects must be taken into account when carrying out experiments such as ultracentrifugation, gel chromatography, viscometry, osmometry, electrophoresis, and other procedures which provide data for the size and charge characteristics of humic substances. Intermolecular charge interactions tend to disappear as the concentrations of polyelectrolytes in solution tend to zero, and so the results obtained from experimentation are often extrapolated to zero concentration. This procedure is usually not practical. The difficulties can be overcome by suppressing the charge effects through the addition of a neutral salt as background electrolyte. Addition of background electrolyte also alters the intramolecular charge interactions, and this decreases the molecular dimensions. Hence, if large amounts of neutral salts are added, molecular shrinkage and solvent exclusion may lead to salting out (and precipitation) of the macromolecules. Suppression of charge will increase the possibilities for the formation of molecular associations, either between the humic macromolecules themselves, or with other molecules or substances (see Swift[81]).

Swift[81] has discussed the relevance of number-average, weight-average, and z-average molecular weight values for humic substances. For polydisperse systems, such as humic substances, where a wide range of molecular weight values are found, the values for the z-average are greater than for the weight average, and these in turn are greater than the number average values. Because, in addition to molecular size polydispersity, humic substances are also highly inhomogeneous with respect to charge density, functional group content, *etc.*, it is difficult to get meaningful results from working with unfractionated samples.

Hayes and Swift,[1,4] and various contributors to 'Humic Substances II. In Search of Structure',[82] have discussed determinations of molecular weight values of humic substances using ultracentrifugation, depression of freezing point, osmometry, gel chromatography, and viscometry. Wershaw and Aiken[83] have also discussed these methods, and applications of the scattering of electromagnetic radiation for determinations of size and weight data for humic substances.

We consider that ultracentrifugation studies have provided the most reliable of the data we have currently concerning the molecular weights, sizes, and shapes of humic substances. The sedimentation velocity and the sedimentation equilibrium procedures are effective for determinations of molecular weight values. In the former, the rate of sedimentation of solute molecules is measured, and the molecular weight (M) calculated from the equation

$$M = 2RTs/(1 - \bar{V}\rho)D \tag{1}$$

where M is the molecular weight, R is the gas constant, T is the temperature (K), s is the sedimentation coefficient, D is the diffusion coefficient, $\bar{V}$ is the partial specific volume, and ρ is the density. This procedure works effectively only when the solute is relatively mono-disperse, and so it should not be applied to unfractionated samples of soil humic substances.

The sedimentation equilibrium procedure can be used for samples which have a degree of polydispersity. In this

$$M = [2RT/(1 - V\rho)\omega^2]d\ln c/dx^2 \tag{2}$$

where x is the angular velocity, c is the concentration of solute at a distance x from the axis of rotation, and the remaining symbols are as defined for the sedimentation velocity method. Account must be taken in ultracentrifugation studies of the charge effects on the humic macro-molecules. Because the large molecules sediment much more rapidly than the small molecules, the polyanionic humic polyelectrolytes will fall faster than the counter-ions. Conversely the small counter-ions diffuse more rapidly than the macromolecules. There are, therefore, opposing drag and pulling effects. Decreased values for s and increased values for D will give lower molecular weight values. Hence, in order to counter these charge effects, it is necessary to add background electrolyte (such as 0.1–0.2 M KCl).

Ultracentrifugation provides useful information about shapes as well as sizes when use is made of frictional coefficient (f) data. The frictional coefficient of the molecule is given by

$$f = kT/D \tag{3}$$

where k is the Boltzman constant, T is the temperature (K), and D is the diffusion coefficient of the macromolecule in the solvent medium. Deviation from the condensed spherical shape is given by the frictional ratio, f/f_0 or f/f_{min}, where f_0 (or f_{min}) is the frictional coefficient of a condensed sphere occupying the same volume as the molecule under study. Molecules which have the condensed rigid sphere conformation have f/f_0 values of 1.0, and this is characteristic of many proteins. Molecules which are ellipsoidal have values in the range of 1.5–3, and rod-shaped molecules can have values greater than 3.

Cameron *et al.*[37] overcame the polydispersity problem by fractionating the humic acids from a sapric histosol into 11 fractions. The components of the two lowest molecular weight fractions had been isolated using neutral sodium pyrophosphate. A further seven fractions of increasing molecular weights were obtained from extracts in sodium hydroxide at 20 °C, and the two samples of highest molecular weight were isolated in sodium hydroxide solution at 60 °C. Molecular weight values were determined by equilibrium ultracentrifugation, and these ranged from 2.6×10^3 to 1.36×10^6. They also determined frictional ratio values, and these ranged from 1.14 for the smallest molecules to 2.41 for the largest. The relationship between frictional ratio values and molecular weight is shown in Figure 6, and the solid straight line represents the theoretical relationship $f/f_{min} = 0.30\, M^{1/6}$, which is considered to be characteristic of either a random coil, or of a discoid type of structure. It is evident that this line corresponds well to the data obtained for fractions A1 to B6 (of $M = 199\,000$).

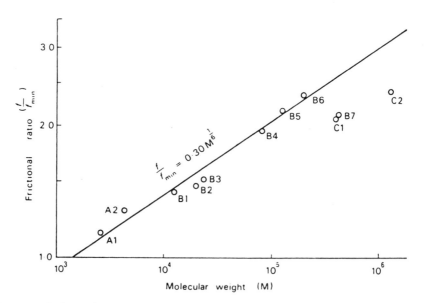

Figure 6 *Relationship between molecular weight and the frictional ratio for humic acids. Experimental data are shown as circles and the straight line is the theoretical derived relationship (from ref. 37)*

Considerations of the modes of formation of humic substances, the irregular arrangements of component molecules in the macromolecular structure, their ion-exchange behaviour, their high water regain, *etc.*, led Cameron *et al.* to the conclusion that the random coil is the most likely conformation which the humic macromolecules will adopt in solution (see also below). The deviations of the higher molecular weight samples (B7, $M = 412\,000$; C1, $M = 408\,000$; and C2, $M = 1.36 \times 10^6$) from the linear relationship (Figure 6) was interpreted as an indication of increased chain branching, and/or cross linking.

Radius of gyration data (R_G) can also give indications of molecular size. Fraction A1, for example, had an R_G value of 1.5 nm, while the values for B6, C1, and C2 were 13.2, 15.0, and 25.5 nm, respectively. In comparison, the proteins serum albumin ($M = 66\,000$), catalase ($M = 225\,000$), and urease ($M = 480\,000$) had R_G values of 3.0, 4.0, and 9.0 nm, respectively. These values show that the humic substances in solution have three-dimensional structures that are much more expanded than those of proteins. This concept of shape has many important practical implications, as outlined in Section 6, because it is possible to make sensible predictions of how dimensions and shapes will change as the environment of the macromolecule is altered.

Molecular Shapes and Sizes of Soil Polysaccharides

There are few accurate data available for the sizes and shapes of soil polysaccharides, and this can be attributed to the problems which must be faced in any attempt to isolate from soil a polysaccharide that is homogeneous with respect to size. (Charge polydispersity is easier to deal with for soil polysaccharides.) It is clear, however, that some soil polysaccharides have high molecular weight values, and there are reports of fractions with values ranging from a few thousand to several hundred thousand Daltons.[23,24,41]

The shapes of the polysaccharides will be determined primarily by the nature (α or β) of the glycosidic linkages.[25] For example, the glucose units of cellulose are linked through the β-$(1 \rightarrow 4)$ configuration. This allows the chain of glucose molecules (in the chair conformation) to assume a linear helical structure which leads to extensive hydrogen bonding between the component molecules in adjacent strands. That explains the very low solubility of cellulose in water, and the fact that it has some degree of resistance to biological breakdown. In contrast, the α-$(1 \rightarrow 4)$ configuration gives rise to a more open structure (as in starch), and the glycosidic linkages are more readily cleaved.

6 Concepts of Structure and Relevance to Function

In their considerations of the structures of humic substances, Hayes *et al.*[58] took account of the regularity which gives to proteins their well defined primary, secondary, tertiary, and quaternary structures. They discussed how the lack of genetic control in the synthesis of humic substances would make it unlikely that such macromolecules will have the regularities in the arrangements of the component molecules to give well ordered structures, such as those that are representative of biological polymers. However, as described in Sections 2 and 5, the macromolecules can be fractionated on the basis of size (and in principle on the basis of charge density) differences, and such fractionation facilitates determinations of molecular sizes. Although all of the molecules which compose humic macromolecules are not known unambiguously, it is possible, in theory, to deduce from the products of degradation what these molecules are likely to be.

There is, of course, biological control of the synthesis of soil poly-saccharides. Hence it will be relatively straightforward to carry out a full structural study when a single macromolecular species is isolated from the soil. Our concepts of how soil polysaccharides function in the soil environment are greatly facilitated by the availability for research of a variety of polysaccharides of known compositions and structures. These serve as useful models which allow us to investigate the mechanisms by which the different kinds of polysaccharides that occur in soil can react with the mineral and with the other organic components of the soil.

Due emphasis has been placed on polydispersity in composition, and on the molecules and functional groups which compose humic substances. It is inevitable that such polydispersity should exist because of the heterogeneity of the substrates that give rise to the macromolecules, and the processes involved in their genesis. Because it is unlikely that there are two humic molecules in any batch that are exactly the same, it would be pointless to try to interpret the data from degradation reactions and from spectroscopic studies in terms of accurate macromolecular structures. However, it is not necessary to know the exact structures in order to have a good understanding of composition, and to be able to deduce the mechanisms and processes of reactions and interactions which involve humic macro-molecules in soil and water environments. In fact, our awareness at this time of composition and structure relevant to humic substances allows us a degree of understanding of the processes involved in their binding of metals and of anthropogenic organic chemicals, in their adsorption to clays and to hydr(oxide) minerals, and in their reactions with compounds used in water disinfection processes, such as chlorine and ozone. Nevertheless, unambiguous interpretations of mechanisms of interactions involving humic substances will not be possible until there is a better awareness of the component molecules, of the linkages between these molecules, and of the juxtapositions and spacings of the reactive functional groups in the macro-molecular structures.

The information which we have suggests that aromatic structures contribute significantly to the reactivities of the macromolecules. There is evidence to suggest that 35–40% of the composition of soil humic acids is made up of aromatic structures,[51] and it is very likely that these are single ring compounds. Such structures are thought to be major contributors to the binding of guest (aromatic) molecules (especially through charge-transfer processes) by humic substances.

The hydrogens in three to five of the aromatic ring positions may be replaced by substituent groups, and these groups (in addition to hydroxyl groups and methoxy structures) may often include aliphatic components linking the aromatic moieties and extending the structures. There is evidence too for aldehyde and keto functional groups associated with some of the aromatic nuclei, and for phenylpropane (3-carbon chains attached to the aromatic rings) units, or structures derived from such units, and for hyroxyl and methoxy substituents on the 3-, 4-, and sometimes the 5-ring positions. Some such structures would indicate origins in lignins. Aromatic structures bearing aliphatic hydrocarbon substituents, and with hydroxyl and methoxy groups in the 3- and 5-ring positions, suggest that some of the components found in degradation digests have their origins in the 'skeletal structures' and/or in the metabolic products of microorganisms. There is evidence to indicate that ether functional groups link aromatic units, and it is inevitable that aromatic–aliphatic ether linkages are also present. It is not necessary to consider that all of the aliphatic groups found in the digests of degradation reactions of humic substances are involved in linking aromatic structures. Some are likely to be substituents on single aromatic rings. Long-chain hydrocarbon structures, possibly with origins in plant cuticles and in algal metabolites, can be attached to aromatic rings, but most are likely to be present as 'impurities' held by van der Waals forces to the humic macromolecules. Some of the hydrocarbon structures are olefinic, and these are cleaved to carboxylic acid structures during the course of chemical oxidation. [Such structures would account for the dicarboxylic acids found in digests of oxidative degradation reactions (see Section 4).]

Fatty acids found in digests of degradations of humic substances seem likely to be released from esters of phenols, and/or from other hydroxyl groups in the 'backbone' structures. These may also be components of waxes, and of suberins associated with the humic substances. The fatty acid and the hydrocarbon moieties contribute to the hydrophobic properties which humic substances display in some circumstances.

Sugar and amino acid residues, which may be components of oligosaccharides and polysaccharides, and of peptide structures, can be linked to the humic acid 'backbone' through covalent links. Amino acids invariably contribute more to the structures than do sugars. However, the combined contributions to humic structures of sugars and amino acids are generally less than 10% of the masses of humic macromolecules.

Titration data show that the acid groups in humic and fulvic acids provide a continuum of dissociable protons representing a range of acid strengths, from the strong to the very weak. The strongest acids are carboxylic, and the strongest of these are activated by appropriate adjacent functional groups. Phenolic hydroxyls contribute to the acidity, and more to that of humic acids than fulvic acids. It is likely that their contribution to the total acidity will be greatest for newly formed humic substances, and especially when these have their origins in the lignified components of plants. As humification proceeds, the phenols are oxidized (see Section 3), and eventually carboxylic acids are formed. Again, neighbouring substituents influence dissociations of the phenolic hydroxyls, and some phenolic functionalities can be relatively strongly acidic. Benzene carboxylic acids, having (phenolic) hydroxyl groups *ortho* or/and *para* to the carboxyl (see Section 4) are relatively strong acids, and such structural arrangements provide powerful ligands for complexation reactions. Enols, and other weakly dissociable groups also contribute to the charge characteristics under alkaline conditions.

The information in the preceding section outlines the concepts we have of the shapes and sizes of humic substances based on results from one extensive study of humic acids isolated from one soil. That study provides good evidence to indicate that high molecular weight humic acids can be expected to take up the random coil conformation in solution. Negative charges are distributed along the lengths of the molecular strands, and the strands coil randomly with respect to time and space. The shapes which result are roughly spherical. There is a Gaussian distribution of molecular masses, with the mass densities greatest at the centre and decreasing to zero at the outer edges. The shapes of the Gaussian mass distribution will vary, depending on whether the molecules are tightly or loosely coiled, and this in turn will depend on the extent of solvent penetration, the charge densities, the degrees of dissociation, the nature of the counter-ions, and the crosslinking.[1,4,81] The evidence we have suggests that the extents of crosslinking are small.

A high degree of dissociation of the component acid groups is necessary for humic acids to dissolve. Considerable energy is needed to overcome the forces holding the macromolecules together in the solid state. As acid is added to the dissolved macromolecules, H^+-exchange takes place. This allows the formation of intermolecular and of intramolecular hydrogen bonding processes causing the structures to shrink, and water to be excluded from each macromolecular matrix. Eventually precipitation takes place. Similar effects result when divalent and polyvalent cations form bridges within and between the strands. As drying takes place, the polar groups will orientate towards the interiors of the structures exposing to the exterior the more hydrophobic moieties. That would explain the difficulty of rewetting dried humic acids.[1,4]

Although soil fulvic acids have some of the characteristics of humic acids, they have significant variations as well. Fulvic acids are less aromatic than humic acids (and can contain as little as 25% aromatic components in some instances), and they are smaller, more polar, and more highly charged. Repulsions between the more closely arranged negative charges on the strands cause the macromolecules to have conformations in solution that are more linear rather than randomly coiled (or spherical). Because fulvic acids do not precipitate when acidified, or in the presence of divalent metals, they would be expected to be readily removed from soils in drainage waters. The colour of waters from upland (acid) peats is, for example, testimony to the loss of water-soluble humic components. It would seem that the hydrogen bonding and van der Waals forces between the soluble fulvic and the insoluble humic acids are not sufficient to hold both types of macromolecules together. (Losses in solution are not confined to fulvic acids, and all drainage waters will also contain some humic acids in solution. However, the solution concentrations are unlikely to be more than 10–20% of those for the fulvic acids, and the molecular weights and dimensions of the humic substances in solution are very much lower than the number average values for those in the solid or gel states in the soil.) Losses of fulvic acids from mineral soils are less extensive than those from upland peats. Humic substances in mineral soils are held by the inorganic colloids, and/or are attached to each other by divalent and polyvalent cation bridges which inhibit their dissolution in the soil solution.

For sorption or other interactions to take place (between sorptive chemicals or metals and humic substances) within the humic macromolecular matrix, it is necessary for the interacting species to diffuse into the matrix. The presence or the absence of water in the macromolecular structures will determine how the charge characteristics are expressed, as well as the flexibilities of the components of the macromolecules, and their abilities to assume arrangements that will allow component structures and functional groups to interact with the guest (or sorptive) species.

The interactions between the soil inorganic colloids and humic substances and polysaccharides is vital for the formation and stabilization of soil aggregates. There is still a gap in our knowledge about the ways in which solid-state humic substances interact with soil clays. It is clear, however, that some humic substances are strongly held by the soil mineral colloids. These resist extraction in aqueous solvents, and are considered to be humin materials. However, some so-called humins can be dissolved in DMSO–HCl[4] and in methyl isobutyl ketone.[84] When removed from the environment of the inorganic colloids, some of these 'humins' have properties similar to those of the more classical humic acids and fulvic acids, and those isolated in DMSO–HCl have a significant carbohydrate content. There is evidence to show that the presence of carbohydrate enhances the abilities of humic acids to stabilize soil aggregates,[85] and some of the so-called humin materials in association with soil clays might be no more than an effective combination of the two most important colloids in the soil to give a product (the stable aggregate) that is essential for soil fertility.

References

1. M.H.B. Hayes and R.S. Swift, in 'The Chemistry of Soil Constituents', eds. D.J. Greenland and M.H.B. Hayes, Wiley, Chichester, 1978, Chap. 3, p. 179.

2. J.W. Parsons and J.Tinsley, *Soil Sci.*, 1961, **92**, 46.
3. D.J. Greenland and J.M. Oades in 'Soil Components', ed. J.E. Gieseking, Springer-Verlag, Berlin, 1975, Vol. 1, p. 213.
4. M.H.B. Hayes and R.S. Swift, in 'Soil Colloids and Their Associations in Aggregates', eds. M.F. DeBoodt, M.H.B. Hayes and A. Herbillon, Plenum, New York, 1990, Chap. 10, p. 245.
5. W. Flaig, in 'Humic Substances III. Interactions with Metals, Minerals and Organic Chemicals', eds. P. MacCarthy, M.H.B. Hayes, R.L. Malcolm and R.S. Swift, Wiley, Chichester, in the press, 1993.
6. 'Soil Colloids and Their Associations in Aggregates', eds. M.F. DeBoodt, M.H.B. Hayes and A. Herbillon, Plenum, New York, 1990.
7. 'Minerals in Soil Environments', eds. J.B. Dixon and S.B. Weed, Soil Science Society of America, Madison, Wisconsin, 1977.
8. M.H.B. Hayes, in 'Humic Substances in Soil Sediment, and Water', eds. G.R. Aiken, D.M. McKnight, R.L. Wershaw and P. MacCarthy, Wiley, New York, 1985, Chap. 13, p. 329.
9. D.C. Whitehead and J. Tinsley, *Soil Sci.*, 1964, **97**, 34.
10. F.J. Stevenson, 'Humus Chemistry', Wiley, New York, 1982.
11. R.W. Taft, D. Gurka, L. Joris, R. Von Schleyer and J.W. Rakshys, *J. Am. Chem. Soc.*, 1969, **91**, 4801.
12. J.H. Hildebrand and R.L. Scott, 'Solubility of Non-Electrolytes', 3rd. edn., Reinhold, New York, 1951.
13. J.H. Hildebrand, J.N. Prausnitz and R.L. Scott, 'Regular and Related Solutions', Von Nostrand-Reinhold, New York, 1970.
14. A.F.M. Barton, *Chem. Rev.*, 1975, **75**, 731.
15. L.R. Snyder, in 'Techniques of Organic Chemistry. Vol XII. Separation and Purification', 3rd. edn., Wiley, New York, 1978, p. 25.
16. A.H. Sinclair and J. Tinsley, *J. Soil Sci.*, 1981, **32**, 103.
17. J.M. Bremner and H. Lees, *J. Agric. Sci.*, 1949, **39**, 274.
18. F.K. Achard, *Grell's Chem. Ann.*, 1986, **2**, 391.
19. M.H.B. Hayes, R.S. Swift, R.E. Wardle and J.K. Brown, *Geoderma*, 1975, **13**, 231.
20. L.N. Alexandrova, *Soviet Soil Sci.*, 1960, **2**, 190.
21. J.M. Bremner, *J. Soil Sci.*, 1950, **1**, 198.
22. N.C. Mehta, P. Dubach and H. Deuel, *Adv. Carbohydr. Chem. Biochem.*, 1961, **16**, 335.
23. G.D. Swincer, J.M. Oades and D.J. Greenland, *Aust. J. Soil Res.*, 1968, **6**, 211.
24. M.V. Cheshire, 'Nature and Origin of Carbohydrates in Soils', Academic Press, London, 1979.
25. M.V. Cheshire and M.H.B. Hayes, in 'Soil Colloids and Their Associations in Aggregates', eds. M.F. DeBoodt, M.H.B. Hayes and A. Herbillon, 1990, Chap. 11, p. 307.
26. M.J. Hausler, 'Studies of the Interactions of Some Imidazolinone Herbicides with Whole Soils, Oxyhydroxides, and with Natural and Synthetic Humic Acids'. Ph.D Thesis, University of Birmingham, 1966.
27. S.A. Barker, M.H.B. Hayes, R.G. Simmonds and M. Stacey, *Carbohydr. Res.*, 1967, **1**, 555.
28. M.V. Cheshire, C.M. Mundie, J.M. Bracewell, G.W. Robertson, J.D. Russell and A.R. Fraser, *J. Soil Sci.*, 1983, **34**, 539.

29. R.S. Swift, in 'Humic Substances in Soil, Sediment, and Water', eds. G.R. Aiken, D.M. McKnight, R.L. Wershaw and P. MacCarthy, Wiley, New York, 1985, Chap. 15, p. 387.

30. G.R. Aiken, in 'Humic Substances in Soil, Sediment, and Water', eds. G.R. Aiken, D.M. McKnight, R.L. Wershaw and P. MacCarthy, Wiley, New York, 1985, Chap. 14, p. 363.

31. A.M. Posner, *J. Soil Sci.*, 1966, **17**, 65.

32. C. Tanford, 'Physical Chemistry of Macromolecules', Wiley, New York, 1961.

33. L. Fischer, in 'Laboratory Techniques in Biochemistry, and Molecular Biology', eds. T.S. Work and E. Work, North Holland, Amsterdam, 1969, Vol. 1, Part II, p. 157.

34. P. Dubach, N.C. Mehta, T. Jakab, F. Martin and N. Roulet, *Geochim. Cosmochim. Acta*, 1964, **28**, 1567.

35. R.S. Swift and A.M. Posner, *J. Soil Sci.*, 1971, **22**, 237.

36. M.G.E. De Nobili, E. Gjessing and P. Sequi, in 'Humic Substances II. In Search of Structure', eds. M.H.B. Hayes, P. MacCarthy, R.L. Malcolm and R.S. Swift, Wiley, Chichester, 1989, p. 561.

37. R.S. Cameron, B.K. Thornton, R.S. Swift and A.M. Posner, *J. Soil Sci.*, 1972, **23**, 342.

38. D.S. Orlov and E. Yu Milanovsky, in 'Volunteered Papers', 2nd Int. Conf. Int. Humic Substances Society (Birmingham, 1984), eds. M.H.B. Hayes and R.S. Swift, University of Birmingham, 1985, p. 42.

39. D. Rickwood, 'Centrifugation: A Practical Approach', Information Retrieval, 1978.

40. M.H.B. Hayes, J.E. Dawson, J.L. Mortensen and C.E. Clapp, in 'Volunteered Papers', 2nd Int. Conf. Int. Humic Substances Society (Birmingham, 1984), eds. M.H.B. Hayes and R.S. Swift, University of Birmingham, 1985, p. 31.

41. P. Finch, M.H.B. Hayes and M. Stacey, *Int. Soc. Soil Sci. Trans. Commun. IV and VI (Aberdeen, 1966)*, 1967, p. 19.

42. C.E. Clapp, 'High Molecular Weight Water-soluble Muck; Isolation and Determination of Constituent Sugars of a Borate-forming Polysaccharide Employing Electrophoresis Techniques', Ph.D Thesis, Cornell University, 1956.

43. C.E. Clapp, J.E. Dawson and M.H.B. Hayes, in 'Proc. Int. Symp. Peat Agric. Horticulture', ed. K.M. Schallinger, Special Publication No. 205, Institute of Soils and Water, Division of Scientific Publications, Bet Dagan, Israel, 1979, p. 153.

44. M. H. Hubacher, *Anal. Chem.*, 1949, **21**, 945.

45. M. Schnitzer and U.C. Gupta, *Soil Sci. Soc. Am. Proc.*, 1965, **29**, 274.

46. M. Schnitzer and S.I.M. Skinner, *Soil Sci.*, 1966, **101**, 120.

47. W. Flaig, H. Beutelspacher and E. Reitz, in 'Soil Components', ed. J.E. Gieseking, Springer-Verlag, Berlin, 1975, Vol. 1, p. 1.

48. R.L. Wershaw, in 'Humic Substances in Soil, Sediment, and Water', eds. G.R. Aiken, D.M. McKnight, R.L. Wershaw and P. MacCarthy, Wiley, New York, 1985, Chap. 22, p. 561.

49. C. Steelink, R.L. Wershaw, K.A. Thorn and M.A. Wilson, in 'Humic Substances II. In Search of Structure', eds. M.H.B. Hayes, P. MacCarthy, R.L. Malcolm and R.S. Swift, Wiley, Chichester, 1989, Chap. 10. p. 281.

50. M.A. Wilson, in 'Humic Substances II. In Search of Structure', eds. M.H.B. Hayes, P. MacCarthy, R.L. Malcolm and R.S. Swift, Wiley, Chichester, 1989, Chap. 11, p. 281.

51. R.L. Malcolm, in 'Humic Substances II. In Search of Structure', eds. M.H.B. Hayes, P. MacCarthy, R.L. Malcolm and R.S. Swift, Wiley, Chichester, 1989, Chap. 12. p. 339.

52. P. MacCarthy and J.A. Rice, in 'Humic Substances in Soil, Sediment, and Water', eds. G.R. Aiken, D.M. McKnight, R.L. Wershaw and P. MacCarthy, Wiley, New York, 1985, Chap. 21, p. 527.

53. N. Senesi and C. Steelink, in 'Humic Substances II. In Search of Structure', eds. M.H.B. Hayes, P. MacCarthy, R.L. Malcolm and R.S. Swift, Wiley, Chichester, 1989, Chap. 13, p. 373.

54. P.R. Bloom and J.A. Leenheer, in 'Humic Substances II. In Search of Structure', eds. M.H.B. Hayes, P. MacCarthy, R.L. Malcolm and R.S. Swift, Wiley, Chichester, 1989, Chap. 14, p. 409.

55. M.H.B. Hayes, P. MacCarthy, R.L. Malcolm and R.S. Swift, in 'Humic Substances II. In Search of Structure', eds. M.H.B. Hayes, P. MacCarthy, R.L. Malcolm and R.S. Swift, Wiley, Chichester, 1989, Chap. 24, p. 689.

56. J.A. Leenheer and T.I. Noyes, in 'Humic Substances II. In Search of Structure', eds. M.H.B. Hayes, P. MacCarthy, R.L. Malcolm and R.S. Swift, Wiley, Chichester, 1989, Chap. 9, p. 257.

57. R.L. Malcolm, *Environ. Int.*, 1992, **18**, 609.

58. M.H.B. Hayes, P. MacCarthy, R.L. Malcolm and R.S. Swift, in 'Humic Substances II. In Search of Structure', eds. M.H.B. Hayes, P. MacCarthy, R.L. Malcolm and R.S. Swift, Wiley, Chichester, 1989, Chap. 1, p. 3.

59. J.W. Parsons, in 'Humic Substances II. In Search of Structure', eds. M.H.B. Hayes, P. MacCarthy, R.L. Malcolm and R.S. Swift, Wiley, Chichester, 1989, Chap. 4, p. 99.

60. F.J. Stevenson, in 'Humic Substances II. In Search of Structure', eds. M.H.B. Hayes, P. MacCarthy, R.L. Malcolm and R.S. Swift, Wiley, Chichester, 1989, Chap. 5, p. 121.

61. M.H.B. Hayes, in 'Advances in Soil Organic Matter Research: The Impact on Agriculture and the Environment', Special Publication No. 90, ed. W.S. Wilson, The Royal Society of Chemistry, 1991, p. 3.

62. M.V. Cheshire, P.A. Cranwell, C.P. Falshaw, A.J. Floyd and R.D. Haworth, *Tetrahedron*, 1967, **23**, 1669.

63. M.V. Cheshire, P.A. Cranwell and R.D. Haworth, *Tetrahedron*, 1968, **24**, 5155.

64. E.H. Hansen and M. Schnitzer, *Fuel*, 1969, **48**, 41.

65. E.H. Hansen and M. Schnitzer, *Soil Sci. Soc. Am. Proc.*, 1969, **33**, 29.

66. R.D. Haworth, *Soil Sci.*, 1971, **111**, 71.

67. S.M. Griffith and M. Schnitzer, in 'Humic Substances II. In Search of Structure', eds. M.H.B. Hayes, P. MacCarthy, R.L. Malcolm and R.S. Swift, Wiley, Chichester, 1989, Chap. 3, p. 69.

68. M.H.B. Hayes and M.R. O'Callaghan, in 'Humic Substances II. In Search of Structure', eds. M.H.B. Hayes, P. MacCarthy, R.L. Malcolm and R.S. Swift, Wiley, Chichester, 1989, Chap. 6, p. 143.

69. L.A. Heredy and M. B. Neuworth, *Fuel*, 1962, **41**, 211.

70. K. Ouchi, K. Imuta, and Y. Yamashita, *Fuel*, 1965, **44**, 29.

71. K. Ouchi and J.D. Brooks, *Fuel*, 1967, **46**, 367.

72. P.E. Kolattukudy, *J. Cell. Biochem.*, 1993, **S17A**, 12.

73. G. Almendros and J. Sans, *Geoderma*, 1992, **53**, 79.

74. G. Almendros and J. Sans. *Soil Biol. Biochem.*, 1991, **23**, 1147.

75. J.M. Bracewell, K. Haider, S.R. Larter and H.-R. Schulten, in 'Humic Substances II. In Search of Structure', eds. M.H.B. Hayes, P. MacCarthy, R.L. Malcolm and R.S. Swift, Wiley, Chichester, 1989, Chap. 3, p. 69.
76. H.-R. Schulten, B. Plage and M. Schnitzer, *Naturwissenschaften*, 1991, **78**, 311.
77. M.V. Cheshire and S.J. Thompson, *Biochem. J.*, 1972, **129**, 19.
78. M.V. Cheshire, C.M. Mundie and H. Shepherd, *Soil Biol. Biochem.*, 1969, **1**, 117.
79. M.V. Cheshire, C.M. Mundie and H. Shepherd, *J. Soil Sci.*, 1971, **22**, 222.
80. R.S. Swift, 'Humic Substances II. In Search of Structure', eds. M.H.B. Hayes, P. MacCarthy, R.L. Malcolm and R.S. Swift, Wiley, Chichester, 1989, Chap. 16, p. 467.
81. R.S. Swift, in 'Humic Substances II. In Search of Structure', eds. M.H.B. Hayes, P. MacCarthy, R.L. Malcolm and R.S. Swift, Wiley, Chichester, 1989, Chap. 15, p. 449.
82. 'Humic Substances II. In Search of Structure', eds. M.H.B. Hayes, P. MacCarthy, R.L. Malcolm and R.S. Swift, Wiley, Chichester, 1989.
83. R.L. Wershaw and G.R. Aiken, in 'Humic Substances in Soil, Sediment, and Water', eds. G.R. Aiken, D.M. McKnight, R.L. Wershaw and P. MacCarthy, Wiley, New York, 1985, Chap. 19, p. 477.
84. J. Rice and P. MacCarthy, *Sci. Total Environ.*, 1989, **81/82**, 61.
85. R.S. Swift, 'Advances in Soil Organic Matter Research: The Impact on Agriculture and the Environment', Special Publication No. 90, ed. W.S. Wilson, The Royal Society of Chemistry, Cambridge, 1991, p. 153.

Interactions between Contaminants and Naturally Occurring Organic Substances

4

Nature of Interactions between Organic Chemicals and Dissolved Humic Substances and the Influence of Environmental Factors

By Nicola Senesi

ISTITUTO DI CHIMICA AGRARIA, UNIVERSITÀ DI BARI, BARI, ITALY

1 Introduction

Soil and water can be hosts to a substantial quantity of organic chemicals, either applied on purpose or accidentally deposited from a variety of sources. Pesticides are applied to soil directly by aerial or surface treatment. Other chemicals can arrive by wet or dry deposition from the vapour phase, or in forms sorbed to atmospheric particulates (*i.e.* fly ash) originating from waste chemicals and solid waste incineration. Pollutant chemicals can also reach soil and waters *via* municipal and industrial wastes, landfill effluents, and composts of various nature applied to the soil as fertilizers or improvers.

Once on the soil surface, the parent chemicals and their degradation products may be subject to various fates. They may enter the soil or be transported to the aquatic environment by run-off into surface waters and by erosion of contaminated soils into streams. Once in the soil, the chemical may percolate through the top layer, penetrate the unsaturated zone, and eventually reach groundwater in its original form or as a breakdown product. The rate and amount of compound penetration through soil is controlled by several factors including the physical and chemical properties of the compound, the amount of water falling on the field soil, the depth and nature of the soil, and the type and extent of interactions the chemical experiences with the various components it encounters in each phase in the soil system. These components may be inorganic, organic or biological materials that interact differently with organic chemicals depending on their intrinsic properties.

Several studies have suggested that organic chemicals show a greater affinity for organic surfaces than for mineral surfaces. The content and nature of soil organic matter plays a major role in the fate of organic chemicals in soil. Groundwaters underlying soils with low organic content are highly vulnerable to contamination by organic chemicals.

The most ubiquitous natural non-living organic materials in all terrestrial and aquatic environments are humic substances. A number of physical, chemical, biochemical, and photochemical properties (see next section) qualify these as privileged natural compounds in the interaction with organic chemicals. Humic substances may affect the fate of organic chemicals in several ways which include adsorption and partitioning, solubilization, degradation by hydrolysis, and photo-decomposition. All these processes are of fundamental importance in determining the fate of the chemical and its subsequent transfer to groundwater, *i.e.* the rate and amount of chemical penetration through the unsaturated zone. These processes also have important implications in biodegradation and detoxication, bioavailability and phytotoxicity, bioaccumulation, and residue persistence and monitoring of organic chemicals.

The main purpose of this paper is to summarize and discuss molecular and mechanistic aspects of the most important modes of interaction between humic substances and some classes of organic chemicals, with emphasis on pesticides and their degradation products.

2 Soil and Aquatic Humic Substances

Humic substances (HS) comprise a physically and chemically heterogeneous mixture of relatively high molecular weight, yellow to black organic compounds of aliphatic and aromatic nature, formed by secondary synthesis reactions (humification) during the decay process and transformation of biomolecules that originate from dead organisms and microbial activity.[1] These materials are exclusive of undecayed plant and animal tissues, their partial decomposition products, and the soil biomass.

On the basis of their solubility in water at various pHs, HS are divided into two main fractions: (i) humic acids (HA), the portion that is soluble in dilute alkaline solution and is precipitated upon acidification to pH 2; and (ii) fulvic acid (FA), the portion that is soluble at any pH value, even below 2.[1]

Approximately 60–70% of the total soil organic carbon occurs in HS. Estimated levels of soil organic carbon on the earth's surface occurring as HS are 30×10^{14} kg.[1] Dissolved aquatic HS constitute 40–60% of dissolved organic carbon (DOC) and are the largest fraction of natural organic matter (OM) in water.[2] The concentration of HS varies for different natural waters, ranging from the high values determined in coloured rivers and wetlands, to the lowest ones measured in groundwaters and seawaters. The majority of the HS in natural waters are FA.

Average elemental composition and acidic functional group content of soil and aquatic HS are presented in Tables 1 and 2. The carbon content of aquatic HS is greater while that of oxygen is less than in soil HS. C/N ratios for aquatic FA (45–55:1) and HA (18–30:1) are considerably greater

Table 1 *Average values of elemental composition (%) of soil and aquatic humic substances*

Sample		C	H	O	N	P	S	Ash
Soil	FA	48.0	4.5	45.0	1.0	—	0.4	1.2
	HA	56.0	4.5	37.0	1.6	—	0.3	2.4
Lake water	FA	52.0	5.2	39.0	1.3	0.1	1.0	5.0
Groundwater	FA	59.7	5.9	31.6	0.9	0.3	0.6	1.2
	HA	62.1	4.9	23.5	3.2	0.5	1.0	5.1
Seawater	FA	50.0	6.8	36.4	6.4	—	0.5	3.4
River water	FA	51.9	5.0	40.3	1.1	0.2	0.6	1.5
	HA	50.5	4.7	39.6	2.0	—	—	5.0
Wetland water	FA	51.0	4.3	40.2	0.7	0.2	0.4	2.0
	HA	51.2	4.4	40.9	0.6	0.1	0.6	2.0

Table 2 *Average values of major functional group (carboxyl and phenolic hydroxyl) content of soil and aquatic humic substances*

Sample		Carboxyl (meq g^{-1})	Phenolic (meq g^{-1})
Soil	FA	5.2–11.2	0.3–5.7
	HA	1.5–5.7	2.1–5.7
Lake water	FA	5.5–6.2	0.3–0.5
Groundwater	FA	5.1–5.5	1.4–1.6
Seawater	FA	5.5	—
River water	FA	5.5–6.0	1.5
	HA	4.0–4.5	2.0
Wetland water	FA	5.0–5.5	2.5
	HA	4.0–4.5	2.5

than C/N ratios in soil FA (average 20:1) and HA (average 10:1) and in HS from aquatic sediments. Aquatic HS appear, therefore, to be depleted of nitrogen as compared to adjacent soils and sediments. The phenolic OH content of aquatic HS (1–2 meq g^{-1}) is considerably lower than that in soil HS. Soil HA and FA differ in elemental composition, molecular weight and functional group contents.[3] The FA fraction exhibits lower molecular weight and lower levels of total and aromatic carbons. Oxygen and oxygen-containing functional group levels are higher in soil FA than HA.[4] HA contains longer-chain fatty-acid products than FA, suggesting a higher hydrophobicity than FA. ^{13}C NMR spectra indicate that approximately 65% of the carbon in aquatic FA is aliphatic and many of the COOH and

OH groups are attached to aliphatic carbons. Aquatic FA, similarly to soil FA, exhibits low molecular weight (500–2000 daltons), whereas aquatic HA ranges in molecular weight from 2000 to 5000 daltons, or greater, being more polydisperse in nature and colloidal in size.[2]

Humic and fulvic acids cannot be regarded as single chemical entities and described by unique, chemically defined molecular structures. Both are defined by a model structure based on available compositional, structural, functional, and behavioural data and contain the same basic structural units and the same types of reactive functional groups that are common to all the single, variable, and unknown molecules.[1] The macromolecular structure comprises aromatic, phenolic, quinonic, and heterocyclic 'building blocks' that are randomly condensed or linked by aliphatic, oxygen, nitrogen, or sulfur bridges. The macromolecule contains aliphatic, glucidic, aminoacidic, and lipidic surface chains in addition to chemically active functional groups of various natures (mainly carboxylic, but also phenolic and alcoholic hydroxyls, carbonyls, *etc.*) which render the humic polymer acidic. Hydrophilic as well as hydrophobic sites are present.

Humic materials are polydisperse and exhibit polyelectrolytic behaviour in aqueous solution.[1,3,5] Surface activity is an important property of HS promoting interactions, especially with hydrophobic organic chemicals. The amphiphilic character and surface activity increase with increasing pH as —COOH and —OH groups form more hydrophilic sites.[3] High pH and a high concentration of HS depress the surface tension of water. This increases soil wettability and promotes interaction of HS with both hydrophobic and hydrophilic organic chemicals in solution. Surface-active properties of HS assume higher importance in interaction phenomena occurring in aquatic environments. Aquatic FA and HA appear to be more surface active than their terrestrial counterparts. Humic and fulvic acids with their relatively open, flexible, sponge-like structures may trap and even fix organic chemicals that can fit into the voids and holes.[3]

Humic substances also contain relatively high amounts of stable free radicals, probably semiquinones, which can bind certain organic chemicals.[6] The increase in free radical content of HA and FA in aqueous media with increasing pH or with visible-light irradiation enhances markedly their chemical and biochemical reactivity.[6]

3 General Properties of Organic Chemicals and Evidence of Interactions with Humic Substances

The importance of DOC in determining the speciation and long-term environmental and health implications of organic chemicals in terrestrial and aquatic systems has only recently been recognized. A substantial part of the organic chemicals found in aqueous phases may interact with dissolved HS thus significantly affecting their rate of dissolution, volatilization, transfer to sediments, biological uptake and bioaccumulation, or chemical degradation (which may be induced *via* hydrolysis or de-alkylation). The distribution and total mass of a chemical in an ecosystem and its environmental behaviour and fate would, therefore, depend, in part, on the extent of its interaction with HS. The decrease in bioavailability of a chemical interacting with dissolved HS, and the resulting decrease in biological uptake and bioaccumulation, might thus mitigate the biological impact of hydrophobic contaminants. The interaction of organic chemicals with HS would also make difficult their analytical identification and quantitation. HS can also exert a competitive effect on the adsorption of volatile organic chemicals on activated carbon in the treatment of drinking water, thus adversely affecting the efficiency of oxidative water treatment processes.

Organic chemicals that can infiltrate soil and waters include a great variety of pesticides and their biological, chemical and photochemical degradation products; aliphatic and aromatic derivatives of petrol and plastics, such as polynuclear aromatic hydrocarbons (PAH); and phthalic acid diesters (PAE); organic solvents; surfactants; and detergents. These compounds feature a wide variety of physical, chemical, and biological properties and belong to widely differing chemical classes. However, they can be classified into two groups, depending on whether their principal interactions with HS involve specific chemistry or unspecific physical forces. Members of the former group are hydrophilic, ionic or ionizable, basic or acidic compounds; among those of the latter group are hydrophobic, non-ionic, and mainly non-polar compounds.

Cationic Compounds

The most widely studied cationic pesticides are the bipyridilium herbicides, diquat and paraquat, that are generally applied in the form of dibromide and dichloride salts. They exhibit high solubility in water and very low volatility. Diquat and paraquat photodecompose readily when exposed to sun or UV light, but are not photodecomposed when adsorbed onto particulate matter.[7] They are able to form charge-transfer complexes with phenols and other electron-donor molecules.[8] Other cationic pesticides of importance in agriculture are the fungicide phenacridane chloride, the germicide thyamine, and the plant-growth regulator phosphon.

Diquat and paraquat become partly inactivated in highly organic soils.[9,10] Sorption of diquat and paraquat on soil organic matter (SOM) is suggested to be the major factor responsible for the decrease in herbicide activity,[11] although the chemicals are still biologically active toward plants and microorganisms.[12-15] Phenacridane chloride and thyamine have the highest levels of adsorption on SOM, followed by phosphon, diquat and paraquat.[12]

Basic Compounds

Amitrole and *s*-triazines are weakly basic compounds, easily ionizable by protonation, and widely applied to soil as herbicides. Amitrole is soluble in water and behaves chemically as a typical aromatic amine. *s*-Triazines have low water solubility, although this increases at pH values where strong protonation occurs, generally between pH 5.0 and 2.0. Structural modifications of the substituents significantly affect solubility at all pH values. Most of the *s*-triazines can be chemically hydrolysed and partly decomposed by UV and IR radiation, including sunlight, in aqueous systems. They are, in general, relatively volatile. *s*-Triazines possess electron-rich nitrogen atoms that confer electron donor ability, *i.e.* the capacity to interact with electron acceptor molecules, giving rise to electron donor–acceptor (charge-transfer) complexes.

The biological activity and transport in soil of *s*-triazine and triazole herbicides is inversely proportional to the level of SOM due to adsorption phenomena.[12,13,16] Chemical hydrolysis and sunlight photodecomposition of *s*-triazines are also influenced by the presence of DOC.[17-19]

Acidic and Anionic Compounds

Several classes of weakly acidic organic compounds are widely used in agriculture for their herbicidal or insecticidal action. They possess carboxyl or phenolic functional groups capable of ionizing in aqueous media to yield anionic species. Chlorinated aliphatic acids show the highest water solubility and strongest acidity. Substituted phenols, such as dinitrophenols, dinoseb, dalapon, ioxynil, bromoxynil, pentachlorophenol (PCP), and the phenoxyalkanoic acids, such as (2,4-dichlorophenoxy) acetic acid (2,4-D) and (2,4,5-trichlorophenoxy) acetic acid (2,4,5-T) have intermediate to low solubility in water. All these compounds are, however, highly water soluble when applied to soil in the common anionic, alkali salt formulations. With the exception of picloram and phenols, acidic pesticides are substantially non-volatile. Other important acidic herbicides are halogenated benzoic acids chloramben, dicamba, and 2,3,6-trichlorobenzoic acid (TBA), and the pyridine derivative picloram.

The bioactivity and transport of acidic pesticides and their ester formulations vary with the SOM content, even though the adsorption level of these herbicides is much lower than that of cationic and basic pesticides.[10,12] Persistence of acidic pesticides, measured in terms of biological activty, and residual toxicity, is highest in soils rich in OM.[12,13,20] Leachability of picloram, amiben, and PCP is negatively correlated with OM,[20,21] whereas the efficacy of PCP decreases with increase in OM due to adsorption.

Non-ionic Compounds

Non-ionic pesticides do not ionize significantly in aqueous systems and vary widely in their composition, physical properties (such as water solubility), polarity, volatility, and chemical reactivity. Chlorinated hydrocarbon insecticides are among the most widely studied non-ionic pesticides. With the exception of lindane, all these compounds are insoluble in water. 1,1,1-Trichloro-2,2-bis(*p*-chlorophenyl)ethane (DDT) is about ten times more insoluble than the other compounds of this family and is considered immobile in soil systems. Endrin, dieldrin, and aldrin show higher water solubility and are hence slightly mobile in soils. In general the water solubility of DDT and other chlorinated hydrocarbons is considerably greater in the presence of dissolved HS than in their absence. The vapour pressure of chlorinated hydrocarbons varies widely: from low (DDT, endrin and dieldrin), to moderate (toxaphene and aldrin), to high (chlordane and lindane), and very high (heptachlor). Volatilization of DDT from soils and other surfaces is, therefore, almost insignificant. Insecticidal activity, degradation, inactivation, leaching, and volatilization of several chlorinated hydrocarbons, including aldrin, dieldrin, endosulfan, lindane, heptachlor, DDT, toxaphene, and chlordane decrease as the SOM content increases. The most prominent effect occurs in moist soils.[10,22,23] In particular, retention and inactivation of DDT is quite generally correlated with SOM content and especially with the humified fraction.[10] Higher levels of SOM also increase the persistence of DDT, lindane, and aldrin, which is higher in a muck soil than in a mineral soil.[24] Transport of DDT in forest soils has been associated with soil HA and FA fractions.[25] OM is reported to be the principal means of deactivation of DDT as determined by a bioassay technique.[26]

Organophosphates, including malathion and parathion insecticides and the herbicide glyphosate, are more toxic, and exhibit higher water solubility and vapour pressure but lower persistence in soil, than chlorinated hydrocarbons. Malathion and parathion insecticides are readily hydrolysed and biodegraded by microorganisms in soil systems. The insecticidal activity of organophosphates and their adsorption in soils is correlated with SOM content.[13] The bioactivity of phorate decreases with increased SOM content,[27] while diazinon and parathion show the same relationships in moist soils, but not in dry soils.[28] Fonofos is persistent for more than two years in an organic soil and its mobility and persistence in soils is suggested to be partly a function of adsorption on HS.[29,30]

Phenylcarbamates, or carbanilates, such as carbaryl, methiocarb, aldicarb, and carbofuran, generally exhibit low water solubilities and are therefore almost immobile in soil systems. Chlorpropham and propham are readily volatilized from soil systems, but terbutol and carbaryl are not. The chemical reactions carbamates may undergo, include ester and amide hydrolysis, *N*-dealkylation and hydroxylation. Phenylcarbamate herbicides show a lower herbicidal activity in fine soils than in coarse textured ones. This appears to be related to the higher OM content of the former soils.[31] Phytotoxicity of chlorpropham is greatly reduced by OM added to the soil.[11] Vapour losses of propham and chlorpropham from moist soils decrease as the percentage of OM increases.[20]

Substituted urea herbicides, including fenuron, linuron, monuron, diuron, cycluron, and many others, generally exhibit low water solubility, with the exception of fenuron, and low volatility. These compounds can undergo hydrolysis, acylation, and alkylation reactions in soils and show electron-donor properties, being able to form charge-transfer complexes with electron-acceptor molecules. The herbicidal activity, adsorption and transport of several phenylureas in soils decrease as the SOM content increases.[13,32-35] Addition of OM to sandy soil significantly reduces the herbicidal activity of fluometuron and fenuron in growth chamber studies and in field experiments.[12]

Substituted dinitroanilines, including trifluralin, nitralin, benefin, butralin, and profluralin, show very low water solubilities and low volatility, with the exception of trifluralin. All these compounds are relatively immobile in soil systems. The herbicidal activity of substituted dinitroanilines, *i.e.* trifluralin and benefin, is reduced by SOM.[31,36-39] The mobility of these herbicides in muck soils is much less than in mineral soils, thus suggesting a partial adsorption of these compounds by organic soil colloids.[12] In greenhouse experiments, orizalyn is inactivated by SOM at a much lower level than trifluralin and nitralin due to much lower adsorption of orizalyn by SOM.[12]

The phenylamide herbicide diphenamide is moderately water soluble and not volatile. It probably behaves much like the acetanilides in aqueous and soil systems. Movement and bioactivity of diphenamid decreases as the SOM content increases.[32] Similarly, additions of SOM to model soil systems significantly reduces the herbicidal activity of diphenamid.[12]

Thiocarbamate and carbothioate herbicides generally exhibit low water solubility and high vapour pressures, thus being relatively mobile in soil systems. Surface losses are attributed to volatilization because of their high vapour pressures. The most important non-ionic benzonitrile herbicide is dichlobenil, which is slightly soluble in water and has a low vapour pressure; it is therefore relatively immobile in most soils. Of the acetamide herbicides, N,N-diallyl-2-chloroacetamide (CDAA) has a relatively high water solubility and vapour pressure, with respect to propachlor and alachlor. Increasing SOM content results in decreased bioactivity, movement, leaching and volatilization of relatively volatile pesticides, which include thiocarbamates, acetamides, and benzonitriles such as S-ethyl-dipropylthiocarbamate (EPTC), pebulate, cycloate, 2-chloroalkyldiethyl-thiocarbamate (CDEC), CDAA, and dichlobenil.[13,40-45] Higher levels of OM result in lower initial toxicity of many herbicides and lower losses through volatilization. Thus these compounds are slightly less effective but persist longer in fine soils.[12] Movement and adsorption of aldicarb, carbofuran, and pyrazone were shown to be positively correlated to SOM content.[46,47]

Polynuclear aromatic hydrocarbons (PAH) are hazardous widespread contaminants produced in large quantities from anthropogenic sources, including the combustion of fossil fuel, chemical manufacturing, petroleum refining, metallurgical processes, and some coal, oil shale, and tar sand conversion systems. PAH are neutral, non-polar hydrophobic organic molecules, with the hydrophobicity increasing with molecular weight, *i.e.*

with the number of benzene rings in the structure. As they have a hydrophobic nature, adsorption is very important in determining their fate in surface and sub-surface water–soil–sediment systems. Aqueous concentrations, transport, and surface-associated chemical and biological degradation processes of PAH are highly depending on adsorptive/desorptive equilibria with sorbents, *e.g.* HS, present in the system. In particular, liquid-to-solid-phase partitioning can play a significant role in the adsorption phenomena on HS.

Phthalic acid diesters (PAE) are used mainly as plasticizers, and also as pesticide carriers and insect repellents, in dyes, in cosmetics, and in lubricants. PAE are lipophilic or lyophobic liquids of medium viscosity, practically immiscible with water, and have low vapour pressure. The isolation and identification of small amounts of PAE (0.03% dry weight) from soil, sediment, and aqueous HA and FA[48-50] indicate that the lyophilic HS can interact with lyophobic PAE to form water-stable products. Water-soluble FA may therefore mediate mobilization and transport of PAE in soil solutions and aquatic environments.

4 Adsorption Mechanisms

Adsorption is probably the most important mode of interaction of organic chemicals with OM. The effect of adsorption on chemical migration in soil depends on whether adsorption occurs on insoluble, immobile organic fractions such as HA, or on dissolved or suspended, mobile fractions such as FA. Organic matter in the form of HA or FA can, therefore, either hinder or facilitate transport.

Organic compounds can be sorbed by OM through physical–chemical binding by specific mechanisms and forces with varying degrees of strength. These include ionic, hydrogen and covalent bonding, charge transfer or electron donor–acceptor mechanisms, dipole–dipole and van der Waals forces, ligand exchange, and cation and water bridging. Adsorption of non–polar (hydrophobic) organic compounds can also be described in terms of non-specific, hydrophobic, or partitioning processes between soil water and the soil organic phase.

The type and extent of adsorption will depend on the amount and properties of both the chemical and the humic molecule. The various properties of the pollutant molecule, such as number and type of functional groups, acidic or basic character, polarity and polarizability, ionic nature and charge distribution, water solubility, shape, and configuration, will often result in several possible adsorption mechanisms that may operate in combination.

For any given chemical, a sequence of different mechanisms may be responsible for adsorption onto humic substances. The organic molecule may be sorbed initially by sites that provide the strongest binding, followed by progressively weaker sites as the stronger adsorption sites become filled. Once adsorbed, the chemical may be subject to other processes affecting retention. For instance, some chemicals may further react to become covalently and irreversibly bound, while others may become only physically trapped into the humic matrix. Adsorption processes may thus vary from completely reversible to totally irreversible, that is, the adsorbed chemical may be easily desorbed, desorbed with varying degrees of difficulty, or not desorbed at all.

Adsorption Isotherms

The construction and use of adsorption isotherms from equilibrium adsorption data has been employed by numerous researchers to describe adsorption of organic chemicals onto a solid matrix. In general, two equations are used to quantitatively describe the adsorption of the chemical onto OM, the Freundlich equation and the Langmuir equation.[51]

Adsorption of linuron, 2,4-D, picloram, and fonofos onto HA is shown to follow a Freundlich-type isotherm.[30,52,53] Adsorption of cationic compounds, including paraquat and diquat, phosphon, hyamine and chlordimeform on SOM results in L-shaped or Langmuir-type isotherms, characterized by a curvilinear response over all concentrations, which seems to level off at certain adsorption maxima.[12,54] This indicates specific adsorption on homogeneous sites. pH-Dependent adsorption of *s*-triazines by organic soil colloids and HA also shows L-shaped isotherms, indicating that adsorbed species are in equilibrium with species in solution at each pH value.[55–57] Adsorption of chlordimeform and atrazine on HA also has the form of a Langmuir isotherm,[54,58] whereas adsorption of terbutryn on HA fits both isotherms.[59] However, neither the Freundlich nor the Langmuir equation fit the data if the adsorption of an organic chemical onto OM is predominantly due to an ion-exchange mechanism, *e.g.* the adsorption of paraquat on HA.[60] In this instance, only the Rothmund–Kornfeld equation fits the results satisfactorily.

In summary, therefore, carefully controlled adsorption measurements and shapes of related isotherms may provide information on the mechanism involved in the adsorption of organic chemicals onto HS.

Ionic Bonding (Cation Exchange)

Adsorption *via* ionic bonding, or cation exchange, applies only to those pesticides which are in the cationic form in solution or can accept a proton, *i.e.* protonate, and become cationic. This process involves ionized, or easily ionizable, carboxylic and phenolic hydroxyl groups of HS.

Infrared[54,60,61] and potentiometric titration[62] data support ionic binding as the dominant mechanism for adsorption of diquat, paraquat, and chlordimeform by HS. Divalent cationic bipyridilium herbicides can react with two negatively charged sites on HS, *e.g.*, two —COO⁻ groups or a —COO⁻ plus a phenolate ion (Figure 1a). Humic and fulvic acids appear to retain paraquat and diquat at levels that are considerably lower than the

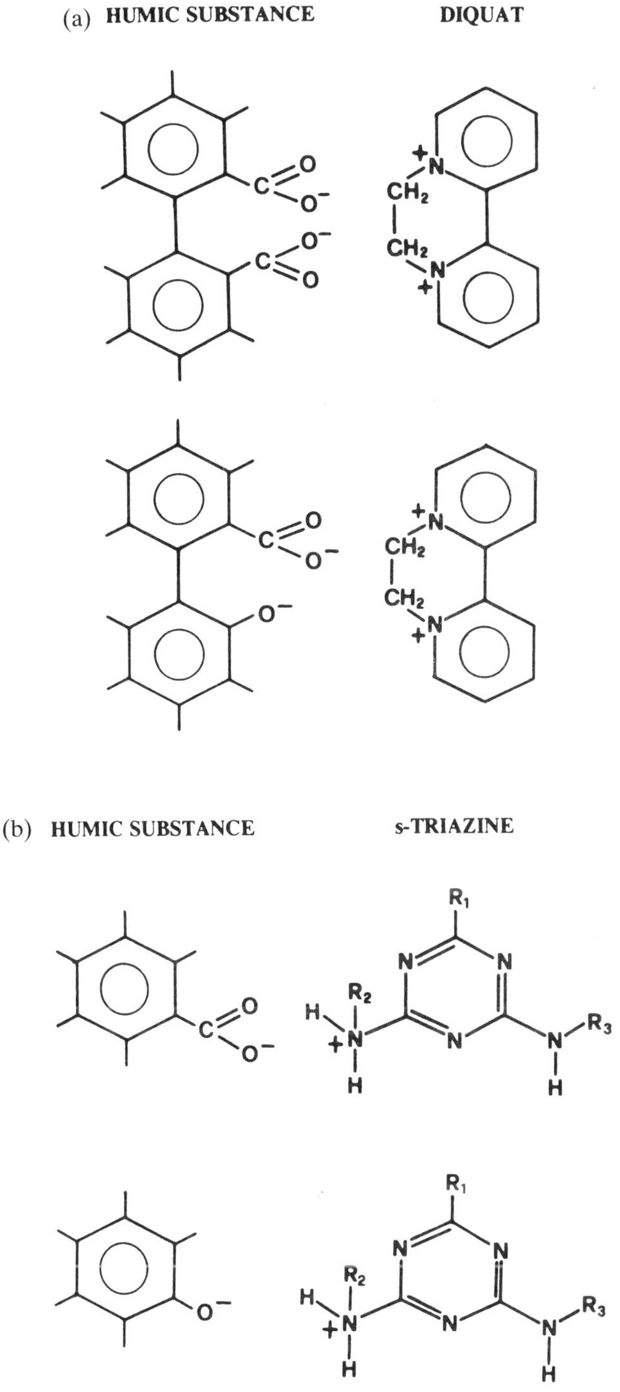

Figure 1 (a) *Model ionic binding between HS and diquat and* (b) *HS and a protonated* s-*triazine*

exchange capacity of the HS.[9] Thus, not all negative sites on HS seem to be positionally available to bind large organic cations, probably because of steric hindrance. Phosphon and phenacridane chloride are also reported to be adsorbed onto SOM through ionic bonding.[12]

Less basic pesticides such as *s*-triazines,[56] amitrole[63] and dimefox[64] may become cationic through protonation depending on their basicity and the pH of the system, which also governs the degree of ionization of acidic groups on the HS. Evidence that the optimum adsorption of *s*-triazines onto organic soils occurs at pH levels close to the pK_a of the herbicide is indicative of ion exchange.[56] However, the pH at the HS surface may be two orders of magnitude lower than that of the liquid phase,[16] thus surface protonation of a basic molecule may occur even though the measured pH of the medium is greater than the pK_a of the adsorbate. IR studies of *s*-triazine–HA systems show that ionic bonding can occur between a protonated secondary amino group of the *s*-triazine and a carboxylate anion and, possibly, a phenolate group of the HA (Figure 1b).[65-72] Differential thermal analysis (DTA) data confirm IR results by showing increased thermal stability of HA upon its interaction with various *s*-triazines.[69,70] Structural features of *s*-triazines, such as the nature of the substituent in the 2-position and of the alkyl groups at the 4- and 6-amino groups are shown to influence their basicity and, hence, reactivity with HS.[73,74] The higher reactivity of simazine with respect to atrazine and prometryne may be related to the smaller steric hindrance of the reactive N—H group of the former herbicide.[66]

Hydrogen Bonding

The presence of numerous oxygen- and nitrogen-containing functional groups on HS renders highly probable the formation of H-bonding for chemicals containing suitable complementary groups, although strong competition with water molecules is expected for such sites. Heat of formation, IR and DTA data suggest the occurrence of one or more H-bonds between HA and *s*-triazines, possibly involving C=O groups of HA and secondary amino groups of *s*-triazines (Figure 2a).[58,65,66,69]

Acidic pesticides such as chlorophenoxyalkanoic acids and esters, asulam, and dicamba can be adsorbed by H-bonding onto humic substances at pH values below their pK_a *via* —COOH, —COOR, and similar groups (Figure 2b).[67,75,76] Hydrogen bonding plays an important role in the adsorption onto HS of several non-ionic polar pesticides including substituted ureas and phenylcarbamates, alachlor, metolachlor, cycloate, malathion, bromacil, and glyphosate[12,63,70,75,77-80] and dialkylphthalates.[81]

Charge Transfer (Electron Donor–Acceptor)

Humic macromolecules contain both electron-deficient structures such as quinones, and electron-rich moieties such as diphenols. This accounts for

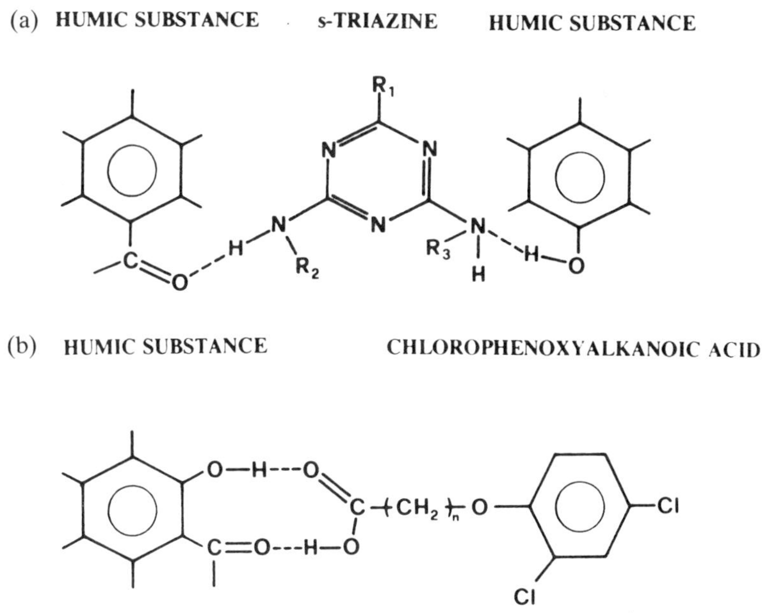

(a) **HUMIC SUBSTANCE** s-**TRIAZINE** **HUMIC SUBSTANCE**

(b) **HUMIC SUBSTANCE** **CHLOROPHENOXYALKANOIC ACID**

Figure 2 (a) *Model hydrogen bonding between HS and a s-triazine and* (b) *HS and a chlorophenoxyalkanoic acid*

their electron-donating and -accepting properties. Electron donor–acceptor mechanisms and formation of charge-transfer complexes will thus be possible with organic chemicals possessing electron-donor or electron-acceptor features.

Substantial spectroscopic evidence exists for the formation of charge-transfer complexes between several HA and *s*-triazines [eqn. (1)], substituted ureas, and amitrole. The shift to lower wavenumber observed in

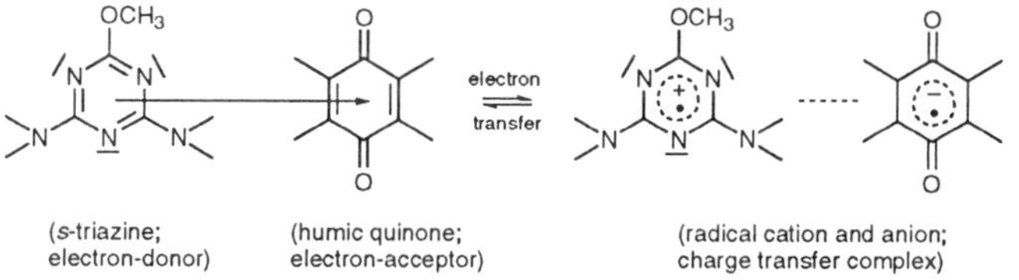

the IR for the out-of-plane deformation vibration of the heterocyclic donor triazine is ascribed to the decreased electron density in the ring resulting from the formation of a complex with electron-deficient molecules of HA.[69,72,82,83] The increase in free-radical concentration measured by electron spin resonance (ESR) in the interaction products of HA with *s*-triazine, substituted ureas, and amitrole is ascribed to the formation of semiquinone free-radical intermediates in the single-electron donor–acceptor transfers occurring between the electron-rich amino or heterocyclic nitrogens of the herbicide and the electron-deficient, quinone-like structures in the HA.[63,69,70,72,77,83] Charge-transfer HA–*s*-triazine systems may also result from the effect of light, which induces an unpairing of electrons involved in the interaction therefore producing the observed increase in the ESR signal (photo-induced transfer).[51]

Structural and chemical properties of both the HA and the organic chemicals may affect the efficiency of formation of the electron donor–acceptor system. The electron-acceptor tendency of the humic molecule seems to be related directly to the quinone content and inversely to the total acidity and the content of carboxyls and phenolic hydroxyls.[69,72] The presence of activating electron donors, such as a methoxyl group on the 2-position of the ring and isopropyl substituents on the amino groups, renders prometone the most efficient electron donor of the *s*-triazines.[69,72] The absence of chlorine atoms on the phenyl ring of fenuron makes this urea a stronger electron donor than ureas containing one or more deactivating electron-withdrawing chlorine atoms on the ring.[72]

IR frequency shifts of the CH out-of-plane bending vibration of the electron acceptors paraquat, diquat, and chloridimeform, observed after interaction with HS, are ascribed to the formation of charge-transfer complexes.[60,61] UV-difference spectroscopy provides evidence of the formation of a charge-transfer complex between electron-donor moieties of HS and the electron-acceptor chloranyl [eqn. (2)].[84] Other important organic

Humic Hydroquinone (Electron Donor) — Chloranil (Electron Acceptor) — Semiquinone Radicals (Electron Donor–Acceptor System)

$$\text{(2)}$$

chemicals such as DDT, dioxins, and polychlorobiphenyls (PCBs) possess electron-accepting properties and may interact with HS by charge-transfer mechanisms similar to those previously described. Such a mechanism has been postulated to be responsible for the photolysis of DDT in the presence of HS.[85]

Covalent Binding

The presence of several types of reactive functional groups in HS affords numerous possibilities for reaction with suitable functional groups of the organic chemical. These reactions are often mediated by chemical, photochemical or enzymatic catalysts and result in the formation of covalent bonds with incorporation of the pollutant or, more likely, its intermediates and products of degradation, into the HS macromolecule.

Acylanilides, phenylcarbamates, phenylamide, phenylureas, dinitroaniline herbicides, nitroaniline fungicides, and organophosphate insecticides, such as parathion and methylparathion, are biodegraded with the release of aromatic amines, such as chloroanilines. In addition, catechols and phenols are released by degradation of several classes of pesticides. These residues form covalent bonds with OM, with or without the intervention of microbial activity. Chloroanilines can be chemically bound to OM by two possible mechanisms including nucleophilic addition to carbonyl and quinone groups and oxidative coupling reactions to phenols [eqns. (3), (4), and (5)]. These reactions may lead to the formation of hydrolysable (*e.g.*

anil, a Schiff base, and anilinoquinone) or non-hydrolysable (*e.g.* heterocyclic rings or ethers) bound forms.[86] The primary amines are thought to interact with HS *via* an initial rapid, reversible formation of amine linkages with the HS carbonyl, followed by a slow, irreversible 1,4 nucleophilic addition to quinone-like rings which is then followed by tautomerization and oxidation to yield an amino-substituted quinone.[87] Slow reaction with HS is also shown by secondary amines. Nucleophilic addition is suggested for the binding to HS of benzidine, α-naphthylamine, and *para*-toluidine.[88]

A marked quenching of free-radical concentration is measured by ESR spectroscopy in the interaction products of soil HA with water-dissolved chlorophenoxyalkanoic acids and esters. This suggests the occurrence of homolytic cross-coupling reactions between indigenous humic free radicals and phenoxy or aryloxy radical intermediates generated photochemically or by chemical or enzymatic catalysis in the initial, partial oxidative degradation of chlorophenoxyalkanoic compounds [eqn. (6)].[76,89] Indigenous inorganic catalysts (*e.g.* copper(II) and iron(III) ions) and residual enzymatic activity (*e.g.* due to phenoloxidases), both of which may mediate chemically or biologically the oxidative degradation of these herbicides, are present in HS.[90] Phenoxy-type radicals may also be generated photochemically from phenoxyalkanoic compounds in the initial oxidation step of the non-biological degradation process they undergo in aqueous solution and in the presence of light and air.[91,92] The coupling reactivity of HA free radicals with chlorophenoxy-derived radicals—assumed to be proportional to the extent of lowering of the residual free-radical concentration in the interaction products—is negatively correlated both with the carboxyl content and COOH to phenolic OH ratio of the HA, and with the number of chlorine atoms on the phenoxy ring of the herbicide. This reactivity would most likely interfere with cross-linking to the HA macromolecule.[76,93]

The significant increase in free-radical concentration and widening of ESR linewidths measured in several HA–*s*-triazine interaction products[69,70,72] confirm the formation of covalent bonds, previously suggested on the basis of the increased adsorption of *s*-triazines on HA observed at high temperatures.[16] These results are attributed to the increased stabilization attained by free electrons on the extended aromatic structures originated by the covalent binding of amino groups of the *s*-triazine to carbonyl and quinone groups of the HA.

Much evidence exists of the enzyme-catalysed incorporation into HA polymers of chloro- and alkyl-substituted aniline residues in the presence of phenoloxidases,[94,95] and of chlorocatechol intermediates of the decomposition of 2,4-D and 2,4,5-T in the presence of peroxidase.[95,96] No evidence is available for the enzymatic binding and incorporation of triazine and phenylurea residues in HS.[97]

Ligand Exchange (Cation Bridge)

Adsorption by ligand-exchange mechanisms involves the displacement of hydration water or other weak ligands partially holding a polyvalent metal ion onto OM by a suitable functional group of the chemical. This group may act as ligand to the metal, with the formation of an inner-sphere complex. *s*-Triazines and anionic pesticides such as picloram are likely to bind HS by this mechanism [eqn. (7)].[98]

| Humic Substance–Cation–Water Bridge | *s*-Triazine | Humic Substance–Cation–*s*-Triazine Bridge | (7) |

Dipole–Dipole and van der Waals Forces

Van der Waals forces are weak, short-range dipolar or induced-dipolar attractions that operate, in addition to stronger binding forces, in all adsorbent–adsorbate interactions. However, these forces assume particular importance in the adsorption of non-ionic and non-polar pesticides. Since van der Waals forces are additive, their contribution increases with the size of the pollutant molecule and with its capacity to adapt to the surface of the humic molecule.

Although scarce experimental evidence is available, van der Waals forces are considered to be actively involved in the adsorption onto HS of large bipyridilium cations,[99] carbaryl and parathion,[100] alachlor and cycloate,[63] benzonitrile and DDT,[101] and several thiocarbamates, carbothioates, and acetanilides,[12] and to represent the principal adsorption mechanism for picloram and 2,4-D.[52,79]

Hydrophobic Adsorption and Partitioning

Hydrophobic adsorption onto the surface, or trapping within interior pores of the humic macromolecular sieve, has been proposed as an important non-specific mechanism for retention of non-ionic, non-polar organic chemicals that interact weakly with water. Hydrophobic active site of HS include aliphatic side chains or lipid portions and aromatic lignin-derived moieties with a high carbon content and bearing a small number of polar groups. Water molecules do not compete effectively with hydrophobic molecules for these sites.

Hydrophobic adsorption by OM is suggested to be important for DDT and other organochlorine insecticides,[101] dialkylphthalates,[49,50] leptophos, methazole, norflurazon, oxadiazinon, butralin, and profluralin,[67] metolachlor,[79] chlorinated hydrocarbons,[101] picloram and dicamba,[52] 2,4-D,[75] parathion,[100] phenylcarbamates,[102] substituted anilines,[12] and PCBs.[103] Hydrophobic association has also been proposed as a possible mechanism for adsorption of the *s*-triazines[104] and the phenylureas[53,105] by SOM.

Hydrophobic retention needs not to be regarded as an active adsorption mechanism and is often considered as a 'partitioning' between water and a non-specific organic phase. In contrast to adsorption, the term partitioning describes a process in which the adsorbate permeates, *i.e.* dissolves, into the network of the organic phase. Thus, partitioning is distinguished from adsorption by the homogeneous, aspecific distribution of the sorbed material throughout the entire volume of the organic phase.[106] Partitioning is modelled as an equilibrium process, similar to that between two immiscible solvents, such as water and an organic solvent like octan-1-ol. Humic substances both in the solid and dissolved phase are thus considered as a non-aqueous solvent into which the hydrophobic pollutant can partition from water.

Thermodynamic arguments have been proposed, but may be too simplified to distinguish between partitioning and adsorption and may lead to erroneous conclusions.[107] The wide variation observed by several authors in the strength of interactions (expressed as partition coefficients) between hydrophobic organic chemicals and HS of different types is attributed to the variation in the chemical and structural composition of the latter.[108] This confirms that strong and 'specific' chemical interactions are involved and should be accounted for in any predictive modelling of hydrophobic association of organic chemicals with HS.[109]

5 Solubilization Effects

Interactions of certain organic chemicals with DOC, mainly FA, can significantly modify their solubility and thus affect their mobility and migration. The physical and chemical properties of the chemical (the solute) and of the HS can markedly affect the solubility of the former. At a given concentration of dissolved HS, relatively water-insoluble, non-ionic organic chemicals, *e.g.* DDT, PCBs, PAHs, *n*-alkanes, PAEs, are most easily solubilized.[110–115] This increase in solubility may be the result of direct adsorption or partitioning of the chemical, or of an overall increase in solvency.[116]

The effect of solubility enhancement decreases with increasing intrinsic water solubility of the chemical.[116] This trend has been observed for relatively water-insoluble higher alkanes in comparison with more water-soluble aromatic compounds[111] and for DDT *versus* lindane.[117] The observed solubility enhancement of the chemical has been described as possibly resulting from a partition-like interaction between solute and OM of high molecular weight. The latter is regarded as a 'microscopic organic phase', similar to a micelle.[112] The effectiveness of this interaction appears to be largely controlled by a number of properties of the humic molecule such as size, polarity, configuration, conformation, composition, total acidity, and surface activity.[112,113,115] The fact that soil HA enhance the water solubility of DDT and PCBs more effectively than do FA and river-water HA is attributed to higher molecular weight, carbon content, and aromaticity in addition to large non-polar volume and lower oxygen and polar group content and hydrophilicity.[116] Low molecular weight OM or highly polar organic chemicals do not exhibit a strong effect.

The magnitude of solubility enhancement of a chemical is also dependent on the concentration of dissolved HS, solution pH, and temperature. The water solubilities of compounds such as DDT, some PCBs and higher alkanes are sensitive to low levels of dissolved HS; those of highly water-soluble chemicals are less so. The solubility enhancement is decreased if the pH is raised for DDT and PCBs.[112,113] At alkaline pH, the hydration and concomitant repulsion of ionized acidic groups of the humic molecule would lead to more expanded conformations and thus decrease the extent of non-polar regions available for the interaction with the hydrophobic chemical. A temperature decrease from 25 to 9 °C increases two-fold the binding constant of DDT to dissolved HS, thus increasing its apparent water solubility.[113]

All these results suggest that the solubility effect exerted by dissolved HS on hydrophobic organic chemicals cannot be completely and satisfactorily explained on the basis of a simple partitioning model of interaction. Specific binding interactions involved in such a complex process should be duly accounted for in any general model by introduction of suitable, additional parameters and mechanisms that can better account for these effects.

6 Hydrolysis Catalysis

Hydrolysis is an important degradative reaction that leads to products with solubility, volatility, reactivity, and other physical and chemical properties different from those of the parent chemical molecule, therefore affecting its behaviour and fate.

Humic substances, particularly in the dissolved phase, are able to exert a catalytic or inhibitory effect in the abiotic hydrolysis of a number of herbicides. In the presence of FA or HA, the acid hydrolysis rates of *n*-alkyl esters of 2,4-D and chloro-*s*-triazines are enhanced,[118,119] whereas that of alkaline hydrolysis of the *n*-octyl ester of 2,4-D (2,4-DOE) is lowered.[120] Sorption on HS has little or no effect on the rates of pH-independent (*i.e.* non-acid–base catalyzed) hydrolysis of some organophosphorothioate esters such as chlorpyrifos and diazinon, halogenated alkenes, aziridine derivatives, and others.[121]

The catalytic effects of HA and FA on the dechlorohydroxylation of the chloro-*s*-triazines simazine, atrazine, and propazine are attributed to specific interaction through H-bonding between the surface carboxylic groups of the humic molecule and the side-chain nitrogens of the triazine [eqn. (8)]. This would enhance the electron-withdrawing effect from the electron-deficient carbon bearing the chlorine atom, thus reducing the activation energy barrier for hydrolytic cleavage of the C—Cl bond and promoting its

(8)

replacement by the weak nucleophile water [eqn. (8)].[58,119] This mechanism explains the correlation established between soil-catalyzed hydrolysis of chloro-*s*-triazines and the amount adsorbed by OM.[19] Successively, a first-order kinetic reaction was measured with respect to herbicide concentration for the hydrolysis of atrazine at pH 4 in an aqueous suspension of HA.[58] The catalytic effect of HA seemed to depend not only on the number of effective acid groups but also on the arrangement of these groups on the HA molecule.[58] Hydrolysis of atrazine in aqueous FA solution at pH $\leqslant 7$ also follows first-order kinetics with respect to the herbicide concentration.[119] Increase in FA concentration results in a higher hydrolysis rate constant and a shortened half-life, but has no effect on the activation energy, which, however, increases with an increase in pH of the reaction mixture. A change in pH of the FA solution would change the types and concentrations of acidic functional groups involved in the hydrolysis of atrazine, which in turn might affect the mechanism of hydrolysis as indicated by the change in the activation energy.[119] Undissociated carboxyl groups of SOM are suggested to be the only catalytically active sites, leading directly to catalysed hydrolysis of the H-bonded atrazine, while phenolic groups have no catalytic effects.[56,122]

The retardation of base-catalysed hydrolysis of 2,4-DOE by solution-phase HS is envisioned in terms of the changes in solution chemistry in the vicinity of the negatively charged (at neutral and alkaline pH values) surfaces of humic molecules.[120] A tentative general mechanism has been proposed for the overall effects of HS on hydrolysis kinetics of hydrophobic chemicals. The model is based on an analogy with micellar catalysis, with only a minor contribution from general acid–base catalysis.[123] The model has been successfully tested with experimental data obtained for base-catalysed hydrolysis of 2,4-DOE[120] and acid-catalysed hydrolysis of atrazine.[58] Similar to anionic surfactants, the effects in the two instances are attributed to electrostatic stabilization of the transition state for the acid catalysis, in which the substrate becomes more positively charged, and to the destabilization of the transition state for base-catalysed hydrolysis, in which the substrate becomes more negatively charged.[123] On the basis of this model, one would expect the effect on base-catalysed hydrolysis to be small for parathion, which associates only weakly with HS, but large for compounds such as DDT that associates strongly with them.[123] In conditions where much higher concentrations of HS are possible, *i.e.* in sewage sludge or in sediment and soil interstitial water, the impact of HS on the hydrolysis kinetics of organic chemicals is predicted to be larger. Hydrolysis reactions of both parathion and DDT are expected, therefore, to be strongly retarded in these environments. An additional, indirect mechanism that may operate on certain reactions is inhibition by HA on hydrolytic enzymes in soils.[124,125] In conclusion, therefore, a general model to predict the effects of HS on the hydrolysis rates of organic chemicals seems not yet to be available, those proposed appearing over-simplified or incomplete.

7 Photosensitization

Sunlight-induced photochemical transformations are important pathways in the abiotic degradation of organic chemicals in the top layer of soil and waters. Photoreactions may modify the physical and chemical properties of pollutants and significantly affect their fate and behaviour in the bulk soil.[126] Light-induced transformations can be classified as direct photolysis processes, initiated by direct absorption of light by the compound, and indirect or sensitized photolysis, involving light absorption by natural 'photosensitizers' or producers of photoreactants.

Humic substances strongly absorb sunlight and may behave, therefore, as initiators of photoreactions, some of which involve reactive secondary products that have important implications in accelerating, increasing, or even determining light-induced transformation of non-absorbing, photochemically-stable organic chemicals. Humic substances principally act as sensitizers or precursors for the production of highly reactive, short-lived species (so-called 'photoreactants') such as the solvated electron, e^-_{aq}; singlet oxygen, $^1O_2$; superoxide anion, $O_2^-\cdot$; peroxy radicals, $RO_2\cdot$; hydrogen peroxide, H_2O_2; and redox-active species including excited states of the humic molecule and humic organic radicals. However, HS can also function as a scavenger of other phototransients such as hydroxy radicals, $HO\cdot$.[127,128]

The formation of e^-_{aq} and humic-derived cation radicals in irradiated solutions of HS is suggested to result from photoejection of an electron from the excited states $^1HS^*$ or $^3HS^*$, according to:

$$^1HS \xrightarrow{\text{light}} {}^1HS^* \rightarrow {}^3HS^* \tag{9}$$

$$^1HS^* \text{ or } {}^3HS^{**} \xrightarrow{1102} [HS^+\cdot + e^-]^* \xrightarrow[-H_2O]{1109 + H_2O} HS^+\cdot + e^-_{aq} \tag{10}$$

The solvated electron is a powerful reductant that reacts rapidly with electronegative xenobiotics such as chlorinated organics, *e.g.* dioxins, that are dehalogenated.[127,128]

The dominant reactions of solvated electrons are, however, with O_2, to form $O_2^-\cdot$ [superoxide ion, eqn. (11)] which, in turn may disproportionate and react with water to form hydrogen peroxide [eqn. (12)]. Excited (triplet-state) HS [$^3HS^*$, eqn. (9)] can transfer energy directly to ground-state oxygen ($^3O_2$) to form singlet oxygen ($^1O_2$), which, in turn, may generate HA cation radicals and the superoxide ion [eqn. (13)]. The latter two species can also be formed directly by reaction of $^3HS^*$ with $^3O_2$. Photochemically derived HS cation radicals may generate organoperoxy cation radicals by reaction with $^3O_2$ [eqn. (14)].

$$e^-_{aq} + {}^3O_2 \rightarrow O^-_2\cdot \tag{11}$$

$$2O^-_2\cdot \to O_2 + O_2^{2-} \xrightarrow{2H_2O} O_2 + H_2O_2 + 2OH^- \qquad (12)$$

$$^3HS^* + {}^3O_2 \to {}^1HS^* + {}^1O_2 \to HS^+\cdot + O^-_2\cdot \qquad (13)$$

$$HS^+\cdot + {}^3O_2 \to HSO_2^+\cdot \qquad (14)$$

Singlet oxygen and superoxide radicals are efficient, but selective photo-reactants for the transformation of various chemicals. The occurrence of singlet oxygen is suggested to be of importance in the elimination of dissociated forms of phenolic compounds (*e.g.* chlorinated phenols), for cyclic dienes and for sulfur compounds.[129] For example, HS are able to catalyse the photo-oxidation to the corresponding sulfoxide of the sulfide group of the thioether insecticides disulfoton, fenthion, methiocarb, and butocarboxim at the soil surface, but not of the sulfur group in methyl-thiotriazenes, thiocarbamates, or dithiolane insecticides.[126] Quantitative kinetic data show that photosensitized oxygenations of 2,5-dimethylfuran and the insecticide disulfoton in air-saturated natural water samples containing HS are at least one order of magnitude faster than those in distilled water.[92,130] The same photo-oxygenated products are obtained in both natural water samples and solutions of soil-derived HS. Furthermore, HS derived peroxy radicals are important photo-oxidants for alkylphenols.[128] Up to half the triplet states of HS are estimated to have energies sufficiently high to photosensitize reactions of many organic chemicals, including polycyclic aromatic hydrocarbons, nitroaromatic compounds, polyenes, and diketones.[130]

The photoexcited humic parent macromolecule can also undergo direct photochemical reactions with some organic chemicals, thanks to the presence of conjugated structures such as keto and quinone groups. The three major possible pathways are energy transfer (or photosensitization), charge transfer, and photoincorporation.[127] The first pathway [see eqn. (15) below] is a type of indirect photolysis with HS in a photo-excited triplet state acting as sensitizer, that transfers energy to previously bound acceptor molecules and so catalyses the degradation of organic chemicals that have low excited-state energies, but cannot absorb sunlight themselves;[127] DDT is such a compound. The process is reversible, that is, HS can also act as acceptors of energy from the excited organic chemical (OP), thus quenching the photodegradation of the latter.

$$^3HS\text{-}donor^* + OP\text{-}acceptor \rightleftharpoons (HS\text{---}OP)^* \rightleftharpoons HS + OP^*$$
$$\downarrow$$
$$OP\text{-photoproducts} \qquad (15)$$

Charge transfer, *i.e.* photoinduced electron transfer reactions, may occur from irradiated HS systems to polyaromatic electron acceptors such as PAHs and paraquat.[51,127] HS can also photosensitize reactions involving hydrogen-atom transfer, which are likely to involve triplet state intermediates.[130] For example, HA, are able to photosensitize reactions involving hydrogen transfer from the nitrogen of aniline to the sensitizer at much higher rates than those observed in the aniline photoreaction in distilled water.[92] In both photoreacting systems azobenzene is a major product resulting from coupling of the anilino free-radical intermediates.[92]

ESR studies have suggested that visible and UV light irradiation of HS may enhance the indigenous free-radical contents of HS[131–133] which are highly susceptible to free-radical mediated interaction with organic chemicals. The free-radical increase observed by ESR in many donor–acceptor systems, such as HA–*s*-triazine and HA–urea herbicides, may be attributable to the unpairing of electrons originating from the formation of charge-transfer complexes under the effect of light.[69,70,77,134] For example, photolysis of atrazine in water is more extensive, although initially retarded, in the presence of 0.01% dissolved FA [eqn. (16)].[135] Fulvic acid

is then able to photosensitize the further degradation of the initially formed 2-hydroxyatrazine to *N*-dealkylated products [eqn. (16)]. UV irradiation (254 nm) experiments conducted with prometryn show a first-order reaction in distilled water and in an HA suspension at pH 3, while second-order reaction rate kinetics are observed in the presence of dissolved HA at pH 6 and 8 and FA at pH 3.6 and 8.[136] An additional, phytotoxic dealkylated product, 4-amino-6-(isopropylamino)-1-triazine, is detected when the photolysis of prometryn is performed in aqueous solutions of HA or FA.[136]

First-order kinetics are found for the photolysis of triphenylborane (TPB) and its decomposition product diphenylborinic acid (DPBA) in the presence of HA and FA.[137] The sensitized enhanced reaction rates for TPB are lower than for DPBA, with HA having a somewhat greater effect. Dissolved oxygen might compete with the organoborates in energy-transfer reactions with the excited HS and thus act as a quenching agent.

Solutions of commercial HA are significantly less active than natural HS in the photosensitization of 3,4-dichloroaniline (3,4-DCA), a metabolite of several important herbicides.[138] This is probably due to trapping of the reactive 3,4-DCA intermediates by the HA. Direct photoincorporation of organic chemicals such as polychlorobenzenes into the humic macromolecule may also occur through radical combination or cycloaddition.[127]

Finally, HS may be able to accelerate, or induce, by photosensitization several photoreactions that certain organic chemicals undergo in the presence of artificial sensitizers such as riboflavin, methylene blue, rose bengal, acetone, and benzophenone, the occurrence of which is unlikely under natural environmental conditions.

8 Conclusions

Organic chemicals interact with HS in the solid and dissolved phases in several ways, influencing the behaviour and fate of the chemicals in soil and waters. In particular, adsorption processes directly or indirectly control all subsequent events affecting organic chemicals by determining how much of the chemical is solubilized and thus moves into the aqueous and gaseous phase, degrades, or is consumed by organisms. Humic substances also catalyse some chemical degradation reactions such as hydrolysis, and influence photodegradation of organic chemicals. These processes may lead to the formation of reaction intermediates having physical and chemical properties distinct from the parent compound and thus exhibiting different behaviour.

Rarely is only a single process involved in the interaction. More often, several reactions with different mechanisms may concur; one, or a few, of which dominate for a given chemical under given conditions. The type and extent of interaction may change with time and ultimately result in irreversible modification of the chemical and biological properties of the pollutant, and its speciation and partition in the soil–water–organism system. This can lead to the formation of intermediates and degradation products having a mobility, toxicity, and persistence widely different from both those of the parent molecule and those of products obtained in the absence of HS.

Complete immobilization of the chemical can occur through stable incorporation in the humic polymer. Alternatively, the chemical may attach itself reversibly to organic fractions and eventually migrate with these, constituting a potential time-delayed source of contamination by a time-delayed release of toxic units. Experimental evidence for at least partial remobilization of pesticide residues has been obtained for dichloroaniline derivatives of propanil, intact methoxychlor, methylparathion, dinitroanilines, and methabenzthiazuron. However, the question remains unresolved and requires further research. Simplistic representations of this interaction, such as depicting it as a partitioning, may not explain the phenomena actually involved in the interaction of most organic chemicals with OM.

Further, adsorption of pesticides onto OM presents problems for the analytical, qualitative and quantitative, determination of pesticide residues in soil and water. It therefore appears necessary to develop new procedures and methods which take these aspects into consideration and lead to a solution.

Procedures which employ enzyme-catalysed polymerization of organic chemicals into HS may be extremely useful for removing or minimizing pollutant concentrations in aquatic and terrestrial environments. In addition, HS may feasibly act as naturally occurring photosensitizers to induce or accelerate photodegradation processes which organic chemicals are known to undergo in the presence of added artificial sensitizers.

Greater understanding of the chemical nature and reactivity of HS, which constitute the major natural compounds interacting with organic chemicals in soil and water, and more knowledge about the mechanisms of their interactions will facilitate the development of a quantitative thermo-dynamic description and kinetic modelling of the behaviour and fate of chemicals in soils and waters. Advanced techniques currently available which seem likely to help in the molecular and mechanistic investigation of these interactions include Fourier transform infrared (FTIR), NMR, ESR, and fluorescence spectroscopies, and differential scanning calorimetry.

Interdisciplinary research efforts are necessary for the establishment of measures to minimize pollution problems concerning soil, water, and micro-organisms and hence the global environment.

References

1. F.J. Stevenson, 'Humus Chemistry: Genesis, Composition, Reactions', Wiley, New York, p. 443.
2. E.M. Thurman, in 'Organic Geochemistry of Natural Waters', Nijhoff-Junk, Dordrecht, 1986, Chap. 10, p. 273.
3. M. Schnitzer, in 'Soil Organic Matter', eds. M. Schnitzer and S.U. Khan, Elsevier, Amsterdam, 1978, p. 1.
4. M. Schnitzer and S.U. Khan, 'Humic Substances in the Environment', Dekker, New York, 1972, p. 327.
5. M.H.B. Hayes and R.S. Swift, in 'The Chemistry of Soil Constituents', eds. D.J. Greenland and M.H.B. Hayes, Wiley, New York, 1978, p. 179.
6. N. Senesi, *Adv. Soil Sci.*, 1990, **14**, 77.
7. A. Calderbank, *Adv. Pest. Contr. Res.*, 1968, **8**, 127.
8. R. Haque and S. Lilley, *J. Agric. Food Chem.*, 1972, **20**, 57.
9. S.U. Khan, *Res. Rev.*, 1974, **52**, 1.
10. S.U. Khan, in 'Soil Organic Matter', eds. M. Schnitzer and S.U. Khan, Elsevier, Amsterdam, 1978, p. 137.
11. D.C. Scott and J.B. Weber, *Soil Sci.*, 1967, **104**, 151.
12. J.B. Weber, *Adv. Chem. Ser.*, 1972, **111**, 55.
13. J.B. Weber and S.B. Weed, *J. Series North Carolina State Univ. Agric. Experim. Station,* Raleigh, NC, 1974, **4087**, 223.

14. B.V. Tucker, D.E. Pack and J.N. Ospenson, *J. Agric. Food Chem.*, 1967, **15**, 1005.
15. B.V. Tucker, D.E. Pack, J.N. Ospenson, A. Omid and W.D. Thomas, Jr., *Weed Sci.*, 1969, **17**, 448.
16. M.H.B. Hayes, *Res. Rev.*, 1970, **32**, 131.
17. D.E. Armstrong, G. Chesters and R.F. Harris, *Soil Sci. Soc. Am. Proc.*, 1967, **31**, 61.
18. L.S. Jordan, W.J. Farmer, J.R. Goodin and B.E. Day, *Res. Rev.*, 1970, **32**, 267.
19. N. Burkhard and J.A. Guth, *Pest. Sci.*, 1981, **12**, 45.
20. S.B. Weed and J.B. Weber, *J. Series North Carolina State Univ. Agric. Experim. Station,* Raleigh, NC, 1974, **3840**, 39.
21. L. Eliasson, V. Hallman and E. Tolf, *Sver. Skogsvardsforb Tidskr.,* 1969, **67**, 491.
22. M.C. Bowman, M.S. Schechter and R.L. Carter, *J. Agric. Food Chem.*, 1965, **13**, 360.
23. W.J. Farmer, W.F. Spencer, R.A. Shepherd and M.M. Cliath, *J. Environ. Qual.*, 1974, **3**, 343.
24. E.P. Lichtenstein, *J. Agric. Food Chem.*, 1959, **7**, 430.
25. T.M. Ballard, *Soil Sci. Soc. Am. Proc.*, 1971, **25**, 145.
26. J.R. Peterson, R.S. Adams, Jr., and L.K. Cutkamp, *Soil Sci. Soc. Am. Proc.*, 1971, **53**, 72.
27. R.E. Kirk and M.C. Wilson, *J. Econ. Entomol.*, 1960, **53**, 771.
28. C.R. Harris, *J. Econ. Entomol.*, 1966, **59**, 1221.
29. S.U. Khan, H.A. Hamilton and E.C. Hague, *Pest. Sci.*, 1976, **7**, 553.
30. S.U. Khan, *Can. J. Soil Sci.*, 1977, **57**, 9.
31. J.B. Weber, S.B. Weed and T.W. Waldrup, *Weed Sci.*, 1974, **22**, 454.
32. H.D. Dubey, R.E. Sigafus and J.F. Freeman, *Agron. J.*, 1966, **58**, 228.
33. S.R. Obien, R.H. Suchisa and Or. Younge, *Weeds*, 1966, **14**, 105.
34. R.L. Douding and J.F. Freeman, *Weed Sci.*, 1968, **16**, 226.
35. R.G. Nash, *Agron. J.*, 1968, **60**, 177.
36. S.M. Lambert, P.E. Porter and R.H. Schieferstein, *Weeds,* 1965, **13**, 185.
37. R.L. Hollist and C.L. Foy, *Weed Sci.*, 1971, **19**, 11.
38. R.M. Menges and J.L. Hubbard, *Weed Sci.*, 1970, **18**, 244.
39. J.B. Weber, *Proc. Soil Sci. North Carolina*, 1971, **14**, 74.
40. F.M. Ashton and T.J. Sheets, *Weeds*, 1959, **7**, 88.
41. S.G. Fang, P. Theisen and V.H. Freed, *Weeds,* 1961, **9**, 569.
42. C.W. Miller, I.E. Demoranville and A.J. Charig, *Weeds*, 1966, **14**, 296.
43. E. Koren, C.L. Foy and F.M. Ashton, *Weed Sci.*, 1968, **16**, 172.
44. E. Koren, C.L. Foy and F.M. Ashton, *Weed Sci.*, 1969, **17**, 148.
45. G.G. Briggs and J.E. Dawson, *J. Agric. Food Chem.*, 1970, **18**, 97.
46. P. Jamet and M.A. Piedallu, *Phytiatr. Phytopharm.*, 1975, **24**, 279.
47. P. Jamet and M.A. Piedallu, *Weed Res.*, 1975, **15**, 113.
48. G. Ogner and M. Schnitzer, *Science*, 1970, **170**, 317.
49. K. Matsuda and M. Schnitzer, *Bull. Environ. Contam. Toxicol.*, 1971, **6**, 200.
50. S.U. Khan and M. Schnitzer, *Geochim. Cosmochim. Acta*, 1972, **36**, 745.
51. N. Senesi and Y. Chen, in 'Toxic Organic Chemicals in Porous Media', eds. Z. Gerstl, Y. Chen, U. Mingelgrin and B. Yaron, Springer-Verlag, Berlin, 1989, p. 37.
52. S.U. Khan, *Can. J. Soil Sci.*, 1973, **53**, 429.

53. S.U. Khan and R. Mazurkewitch, *Soil Sci.,* 1974, **118**, 339.
54. C. Maqueda, J.L Perez Rodriquez, F. Martin and M.C. Hermosin, *Soil Sci.,* 1983, **136**, 75.
55. J.B. Weber, S.B. Weed and J.A. Best, *J. Agric. Food Chem.,* 1969, **17**, 1075.
56. J.B. Weber, S.B. Weed and T.M. Ward, *Weed Sci.,* 1969, **17**, 417.
57. P. Gaillardon, R. Calvet and M. Tercé, *Weed Res.,* 1977, **17**, 41.
58. G.C. Li and G.T. Felbeck, Jr., *Soil Sci.,* 1972, **113**, 430.
59. R.K. Gupta, S. Raman and K.V. Raman, *J. Indian Soc. Soil Sci.,* 1985, **33**, 255.
60. I.G. Burns, M.H.B. Hayes and M. Stacey, *Weed Res.,* 1973, **13**, 79.
61. S.U. Khan, *J. Environ. Qual.,* 1974, **3**, 202.
62. D.R. Narine and R.D. Guy, *Soil Sci.,* 1982, **133**, 356.
63. N. Senesi, G. Padovano, E. Loffredo and C. Testini, in 'Proceedings of the Second International Conference on Environmental Contamination', CEP Cons. Ltd., Edinburgh, 1986, p. 169.
64. R.E. Grice, M.H.B. Hayes and P.R. Lundie, 'Proceedings of the 7th British Insecticide and Fungicide Conference', 1973, Vol. 11, p. 73.
65. J.D. Sullivan and G.T. Felbeck, *Soil Sci.,* 1968, **106**, 42.
66. R. Turski and A. Steinbrich, *Polish J. Soil Sci.,* 1971, **4**, 120.
67. R.D. Carringer, J.B. Weber and T.J. Monaco, *J. Agric. Food Chem.,* 1975, **23**, 569.
68. N. Senesi and C. Testini, *Soil Sci.,* 1980, **10**, 314.
69. N. Senesi and C. Testini, *Geoderma,* 1982, **28**, 129.
70. N. Senesi and C. Testini, *Ecol. Bull. Stockholm,* 1983, **35**, 477.
71. N. Kalouskova, *J. Environ. Sci. Health B,* 1986, **21**, 251.
72. N. Senesi, C. Testini and T.M. Miano, *Org. Geochem.,* 1987, **11**, 25.
73. J.B. Weber, *Spectrochim. Acta, Part A,* 1967, **23**, 458.
74. J.B. Weber, *Res. Rev.,* 1970, **32**, 93.
75. S.U. Khan, *Environ. Lett.,* 1973, **4**, 141.
76. N. Senesi, T.M. Miano and C. Testini, in 'Chemistry for Protection of the Environment 1985', eds. L. Pawlowski, G. Alaerts and W.J. Lacy, Studies in Environmental Science 29, Elsevier, Amsterdam, 1986, p. 183.
77. N. Senesi and C. Testini, *Pest. Sci.,* 1983, **14**, 79.
78. T.M. Miano, A. Piccolo, G. Celano and N. Senesi, *Sci. Total Environ.,* 1992, **123/124**, 83.
79. J. Kozak, *Soil Sci.,* 1983, **136**, 94.
80. P. Gaillardon, R. Calvet and J.C. Gaudry, *Weed Res.,* 1980, **20**, 201.
81. S.U. Khan, *J. Environ. Sci. Health B,* 1980, **15**, 1071.
82. U. Müller-Wegener, *Sci. Total Environ.,* 1987, **62**, 297.
83. N. Senesi, *Z. Pflanzen. Bodenkd.,* 1981, **144**, 580.
84. M.E. Melcer, M.S. Zalewski, J.P. Hassett and M.A. Brisk, in 'Aquatic Humic Substances. Influence on Fate and Treatment of Pollutants', eds. I.H. Suffet and P. MacCarthy, Advances in Chemistry Series 219, American Chemical Society, Washington, 1989, p. 173.
85. L.L. Miller and R.S. Narang, *Science,* 1970, **169**, 368.
86. T.S. Hsu and R. Bartha, *J. Agric. Food Chem.,* 1976, **24**, 118.
87. G.E. Parris, *Environ. Sci. Technol.,* 1980, **14**, 1099.
88. J.G. Graveel, L.E. Sommers and D.W. Nelson, *Environ. Toxicol. Chem.,* 1985, **4**, 607.
89. N. Senesi, C. Testini and D. Metta, 'Proceedings of the International

Conference on Environmental Contamination' CEP Cons. Ltd., Edinburgh, 1984, p. 96.

90. J.M. Bollag, S.Y. Liu and R.D. Minard, *Soil Sci. Soc. Am. J.*, 1980, **44**, 52.
91. D.G. Crosby, in 'Herbicides: Chemistry, Degradation and Mode of Action', eds. P.C. Kearney and D.D. Kaufmann, Dekker, New York, 1976, Vol. 2, Chap. 18.
92. R.G. Zepp, G.L. Baughman and P.F. Schlotzhauer, *Chemosphere*, 1981, **10**, 109.
93. N. Senesi, T.M. Miano and C. Testini, in 'Current Perspectives in Environmental Biogeochemistry', eds. G. Giovannozzi-Sermanni and P. Nannipieri, CNR-IPRA, Rome, 1987, p. 295.
94. J.M. Bollag, R.D. Minard and S.Y. Liu, *Environ. Sci. Technol.*, 1983, **17**, 72.
95. D.F. Berry and S.A. Boyd, *Soil Biol. Biochem.*, 1985, **17**, 72.
96. D.E. Stott, J.P. Martin, D.D. Focht and K. Haider, *Soil Sci. Soc. Am. J.*, 1983, **47**, 66.
97. J.M. Bollag and M.J. Loll, *Experientia*, 1983, **39**, 1221.
98. D.C. Nearpass, *Soil Sci.*, 1976, **121**, 272.
99. I.G. Burns, M.H.B. Hayes and M. Stacey, *Pest. Sci.*, 1973, **4**, 201.
100. J.A. Leenheer and J.L. Aldrichs, *Soil Sci. Soc. Am. Proc.*, 1971, **35**, 700.
101. R.H. Pierce, C.E. Olney and G.T. Felbeck, *Environ. Lett.*, 1971, **1**, 157.
102. G.G. Briggs, *Nature*, 1969, **223**, 1288.
103. H.J. Strek and J.B. Weber, *Environ. Pollut. A*, 1982, **28**, 291.
104. A. Walker and D.V. Crawford, in 'Isotopes and Radiation in Soil Organic Matter Studies', Food and Agriculture Organization/International Atomic Energy Agency IAEA, Vienna, 1968, p. 91.
105. R.J. Hance, *Weed Res.*, 1965, **5**, 108.
106. C.T. Chiou, P.E. Porter and D.W. Schmedding, *Environ. Sci. Technol.*, 1983, **17**, 227.
107. U. Mingelgrin and Z. Gerstl, *J. Environ. Qual.*, 1983, **12**, 1.
108. W.M. Davis and J.J. Delfino, *Am. Chem. Soc., Div. Environ. Chem.*, 1992, 7.
109. Y.-P. Chin and J.W. Weber, Jr., *Environ. Sci. Technol.*, 1989, **23**, 978.
110. R.L. Wershaw, P.J. Burcar and M.C. Goldberg, *Environ. Sci. Technol.*, 1969, **3**, 271.
111. P.D. Boehm and J. Quinn, *Geochim. Cosmochim. Acta*, 1973, **37**, 2459.
112. C.T. Chiou, R.L. Malcolm, T.I. Brinton and D.E. Kile, *Environ. Sci. Technol.*, 1986, **20**, 502.
113. C.W. Carter and I.H. Suffet, *Environ. Sci. Technol.*, 1982, **16**, 735.
114. P.F. Landrum, S.R. Nihart, B.J. Eadie and W.S. Gardner, *Environ. Sci. Technol.*, 1984, **18**, 187.
115. N. Shinozouka, C. Lee and S. Hayano, *Sci. Total Environ.*, 1987, **62**, 311.
116. D.E. Kile and C.T. Chiou, in 'Aquatic Humic Substances. Influence on Fate and Treatment of Pollutants', eds. I.H. Suffet and P. MacCarthy, Advances in Chemistry Series 219, American Chemical Society, Washington, 1989, p. 131.
117. G. Caron, I.H. Suffet and T. Belton, *Chemosphere*, 1985, **14**, 993.
118. B. Struif, L. Weil and K.E. Quentin, *Vom Wasser*, 1975, **45**, 53.
119. S.U. Khan, *Pest. Sci.*, 1978, **9**, 39.
120. E.M. Perdue and N.L. Wolfe, *Environ. Sci. Technol.*, 1982, **16**, 847.
121. D.L. Macalady, P.G. Tratnyek and N.L. Wolfe, in 'Aquatic Humic Substances. Influence on Fate and Treatment of Pollutants', eds. I.H. Suffet and

P. MacCarthy, Advances in Chemistry Series 219, American Chemical Society, Washington, 1989, p. 323.

122. D.S. Gamble and S.U. Khan, *Can. J. Soil Sci.,* 1985, **65**, 435.
123. E.M. Perdue, in 'Aquatic and Terrestrial Humic Materials', eds. R.F. Christman and E.T. Gjessing, Ann Arbor Science Publishers, Ann Arbor, MI, 1983, p. 441.
124. R. Malini de Almeida, F. Pospisil, K. Vockova and M. Kutacek, *Biol. Plant*, 1980, **22**, 167.
125. F.L. Mulvaney and J.M. Bremner, *Soil Biol. Biochem.*, 1978, **10**, 297.
126. G.C. Miller, V.R. Hebert and W.W. Miller, in 'Reactions and Movement of Organic Chemicals in Soils', eds. B.L. Sawhney and K. Brown, SSSA Special Publication No. 22, Soil Science Society of America, Madison, WI, 1989, p. 99.
127. W.J. Cooper, R.G. Zika, R.G. Petasne and A.M. Fischer, in 'Aquatic Humic Substances. Influence on Fate and Treatment of Pollutants', eds. I.H. Suffet and P. MacCarthy, Advances in Chemistry Series 219, American Chemical Society, Washington, 1989, p. 333.
128. J. Hoigné, B.C. Faust, W.R. Haag, F.E. Scully, Jr. and R.G. Zepp, in 'Aquatic Humic Substances. Influence on Fate and Treatment of Pollutants', eds. I.H. Suffet and P. MacCarthy, Advances in Chemistry Series 219, American Chemical Society, Washington, 1989, p. 363.
129. J. Hoigné, H. Bader and L.H. Nowell, *Am. Chem. Soc., Div. Environ. Chem.*, 1987, **27(1)**, 208.
130. R.G. Zepp, P.F. Schlotzhauer and R.M. Sink, *Environ. Sci. Technol.*, 1985, **19**, 74.
131. N. Senesi and M. Schnitzer, *Soil Sci.*, 1977, **123**, 224.
132. N. Senesi and M. Schnitzer, in 'Environmental Biogeochemistry and Geomicrobiology, Vol. II, The Terrestrial Environment', ed. W.E. Krumbein, Ann Arbor Science Publishers, Ann Arbor, MI, 1978, p. 467.
133. N. Senesi, Y. Chen and M. Schnitzer, *Soil Biol. Biochem.*, 1977, **9**, 397.
134. N. Senesi and C. Testini, *Chemosphere*, 1984, **13**, 461.
135. S.U. Khan and M. Schnitzer, *J. Environ. Sci. Health*, 1978, **3**, 299.
136. S.U. Khan and D.S. Gamble, *J. Agric. Food Chem.*, 1983, **31**, 1099.
137. G.L. Mill and D.J. Carter, *Am. Chem. Soc., Div. Environ. Chem.*, 1987, **27(1)**, 146.
138. G.C. Miller, R. Zisook and R. Zepp, *J. Agric. Food Chem.*, 1980, **28**, 1053.

5

A Unified Approach to the Interaction of Small Molecules with Macrospecies

By Uri Mingelgrin and Zev Gerstl

DEPARTMENT OF PHYSICAL AND ENVIRONMENTAL CHEMISTRY, INSTITUTE OF SOILS AND WATER, VOLCANI CENTER, ARO, PO BOX 6, BET DAGAN 50250, ISRAEL

1 Introduction

The transport, stability, and bioavailability of pollutants in the environment are all strongly affected by their interactions with macrospecies. The term macrospecies is defined here as all entities which are larger than most common organic monomers, including soluble macromolecules, colloidal particles, and immobile solid components of the system of interest.

The sorption of small organic compounds on immobile surfaces retards their transport, diminishes their availability and, more often than not, decreases their susceptibility to biotic degradation. Interaction with soluble macromolecules, on the other hand, can increase the apparent solubility (*i.e.* concentration in the liquid phase) of hydrophobic molecules and hence the mobilities of these molecules.[1-3] For example, Wershaw *et al.*[3] found that in an aqueous solution of sodium humate the apparent solubility of 1,1,1-trichloro-2,2-bis(*p*-chlorophenyl)ethane (DDT) increased by a factor of over 200. Chiou *et al.*[4] demonstrated that dissolved organic carbon from natural sources at concentrations as low as several $mg\,l^{-1}$ was able to increase the apparent solubility of DDT and polychlorinated biphenyls (PCBs). With regard to the mobility of organic chemicals in porous media, sorption on colloidal particles has an intermediate effect between that of sorption on immobile surfaces and complexation with soluble macromolecules. Vinten *et al.*[2] demonstrated that sorption of both apolar (DDT) and ionic (paraquat) pollutants on suspended particles may either enhance or retard their transport through soil. The effect of sorption of organic compounds by suspended particles on their transport depends on the relation between particle size and soil pore size distribution. Factors which affect this relationship (such as ionic strength which controls the tendency of colloids to flocculate) will also determine whether sorption on suspended particles will retard or accelerate transport (see Figure 1). The considerable potential of colloidal organic species and soluble macromolecules to affect

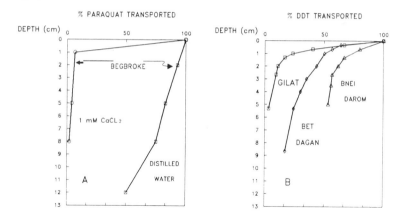

Figure 1 *Vertical transport of DDT and paraquat sorbed on suspended particles in various soils. A. Paraquat sorbed on Li–montmorillonite. B. DDT sorbed on suspended solids derived from a sewage effluent (reproduced from Vinten* et al.[2])

the transport of hydrophobic compounds in soil was demonstrated by Gschwend and Wu[5] and by the computations of Bouchard *et al.*[6]

Despite the contrasting effects sorption on immobile surfaces and complexation with soluble or suspended macrospecies may have on the transport of small molecules, the nature of the interaction between small molecules and macrospecies throughout their wide size range is similar in many respects. This is so, because these interactions are controlled by the same physical principles and parameters regardless of the identity of the specific macrospecies. Interactions of macrospecies, both organic and mineral, of very different chemical and physical character with smaller molecules can thus be understood and predicted using the same few rules and described in a unified way. How the intrinsic properties of the more important classes of macrospecies in the environment influence the sorption–desorption process and the basic similarity of this process regardless of the nature of the macrospecies, are the central themes of this paper.

2 Description of the System at Equilibrium: Mass-action Expression *vs.* Sorption Isotherm

The presence of a sorbate molecule at a sorption site on a surface does not, in general, affect the behaviour of another molecule occupying a non-neighbouring site. As far as pollutants in the environment are concerned, the sorption capacity of the sorbing species is rarely saturated (excepting cases such as spills) and occupation of neighbouring sites on macrospecies is the exception rather than the rule. Thus, as long as the sorbent species contains more than a few sorption sites the interaction between a sorbate molecule and the sorbent is basically the same regardless of the size of the macrospecies. All macrospecies are, as far as a sorbate is concerned, an array of sorption sites, each site being more or less independent. This is true even at higher surface loads, if, as in the derivation of the Langmuir and BET isotherms, interactions between adjacent sorbate molecules are neglected.[7]

The success of the Langmuir and BET isotherms in describing sorption in many systems indicates that viewing macrospecies–small molecule interactions as the sum of interactions between a small molecule and an independent sorption site is a good first-order approximation despite the obvious possiblility of lateral interactions between neighbouring sorbate molecules. Even in the case of multilayer sorption, as described for example by the BET isotherm, the assumption of independence of sorption sites should hold as long as it held for the first sorption layer. A second sorption layer simply means that the nature of the sorption site has changed from that of the original surface to that of a sorbed molecule,[8] while the assumption of independence of sites may remain valid.

'Complexation' with soluble macromolecules and sorption on immobile surfaces or colloids are accordingly identical processes as far as the nature of interaction between sorbent and sorbate is concerned.

The suitability of the Langmuir isotherm for describing the sorption of small molecules on immobile surfaces has been demonstrated for many systems.[9] Less obvious is the applicability of this isotherm to the 'complex-ation' of small molecules with soluble macromolecules. Clapp *et al.* (unpublished data) demonstrated that the complexation of the herbicide napropamide with a soluble humic acid could be satisfactorily described by the Langmuir isotherm (Figure 2).

The following derivation demonstrates why a Langmuir-type equation should be applicable to the complexation of small molecules with soluble macromolecules. At equilibrium, the rates of association and dissociation of the complex should be equal. Then

$$k_1 S = k_2 C(S_m - S) \tag{1}$$

where k_1 and k_2 are the dissociation and association rate constants, respectively, S is the concentration of the complexed molecule, C is the concentration of the uncomplexed small molecule, and S_m is the total concentration of the complexation sites. In turn,

$$S_m = D_c C_h \tag{2}$$

where C_h is the total concentration of the soluble macromolecule and D_c is the density of the available sorption sites, namely the concentration of sorption sites per unit concentration of the macromolecule.

$$S = kCS_m/(1 + kC) \tag{3}$$

where $k = k_2/k_1$.

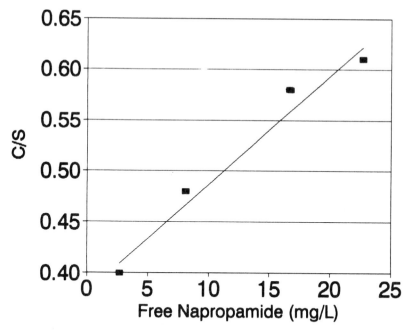

Figure 2 *Langmuir-type isotherm of napropamide complexed with a soluble humic acid* [r² = 0.95; *see equation (7) in text*]. *Humic acid concentration* 2 gl⁻¹ *(Clapp, Liu, Zhang, and Mingelgrin, unpublished data)*

If C_h is defined in units of weight per volume, then

$$S/C_h = C_s \tag{4}$$

where C_s is the concentration of complexed molecules per unit weight of macrospecies. From eqn. (3),

$$C_s = kD_cC/(1 + kC) \tag{5}$$

Equation (5) is precisely the Langmuir isotherm which was originally developed for sorption on surfaces rather than for complexation with soluble macrospecies.

When C is sufficiently small so that $kC \ll 1$, then

$$S/C_hC = kD_c = K_s \tag{6}$$

Equation (6) has the form of a mass-action expression, where K_s is the stability constant (or association constant) of the complex. K_s is often presented as a constant for complexation reactions between small non-ionic molecules and soluble macromolecules.[10] Indeed, at sufficiently low concentrations of the free small molecule (C), K_s is fairly constant. Equation

(3) can be rewritten as

$$C/S = 1/(kS_m) + C/S_m \tag{7}$$

Namely, for a given C_h, C/S should depend linearly on C. The good linear fit of the curve in Figure 2 is in agreement with the above development. The sorption site density (D_c) and the association constant $(K_s = kD_c)$ can be extracted from the slope and intercept of the plot of C/S vs. C [eqn. (7)]. Thus, the applicability of a Langmuir-type isotherm to complexation with soluble macromolecules enables the calculation of the maximum number of small molecules that can be complexed per unit weight of the macromolecule. For example, from Figure 2 the density of the sorption sites on the investigated soluble humic acid for napropamide is $0.18 \, \text{mmol g}^{-1}$ or $48 \, \text{mg g}^{-1}$.

Another outcome of eqns. (4) and (5) is that as C_h increases while maintaining the total concentration of the sorbate $(S + C)$ constant, the quotient $S/(C_hC)$ will approach the constant value K_s [eqn. (6)], since S increases with C_h and C must then decrease to the point that $kC \ll 1$. This is in agreement with the results presented in Figure 3 for the complexation of napropamide by a soluble humic acid.

The assumption of independence of sorption sites can be taken one step

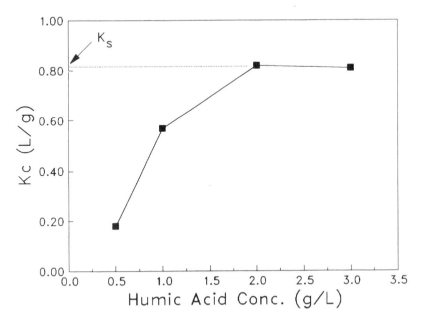

Figure 3 *Complexation ratios* (K_c) *of napropamide with a soluble humic acid at different humic acid concentations.* $K_c = S/(C_hC)$; $K_s = $ *Asymptotic value of* K_c *[see equations (1) and (2) and (6) in text]; total napropamide concentration* $35 \, \text{mg l}^{-1}$ *(Clapp, Liu, Zhang, and Mingelgrin, unpublished data)*

further. If the interacting sites are, to a good approximation, independent and do not affect sorption or complexation on adjacent sites, it should not matter if the sorption sites are connnected to each other or not. Namely, complexation (non-reactive association) between two small molecules should also obey the Langmuir expression and, as will be shown below, indeed they do.

Our perception of the sorption process is such, that an occupied sorption site is not considered an entity distinct from an unoccupied site. Consequently, when an interaction is described by a sorption isotherm the sorbent is assumed to be of a constant concentration. The dependent and independent variables of the isotherm are the amount of sorbate sorbed per unit weight of sorbent and the bulk concentration of the free sorbate, respectively. The change in the fraction of the sorbent which is either free or occupied (*i.e.* the concentration of free or occupied sorption sites) does not appear explicity in the isotherm. This is analogous to describing small sorbate–small sorbent interactions in terms of the concentration of the free molecule (A) and the concentration of the same molecule which is complexed with the second partner (AB). The concentration of the second molecule (B) does not in that case appear in the equation. The independent variable is then the concentration of free molecules (A) and the dependent variable the concentration of the interacting pair (AB).

At equilibrium

$$A + B \rightleftharpoons AB \qquad (8)$$

The standard form used to describe the relation between the concentrations of the species defined in equation (8) is

$$[AB]/[A][B] = K_0 \qquad (9)$$

where the square brackets denote activity and K_0 is a constant. This is a mass-action expression. We can rewrite eqn. (9) as

$$[AB]/[A]([B]_i - [AB]) = K_0 \qquad (10)$$

where $[B]_i$ is the initial, namely total, concentration of B. Rearranging eqn. (10) yields

$$C_s = [AB]/[B]_i = [A]K_0/(1 + [A]K_0) \qquad (11)$$

Equation (11) is analogous to the Langmuir isotherm [eqn. (5)]. C_s denotes the amount of bound A per fixed quantity (*e.g.* unit weight or mole) of B. D_c [eqns. (2) and (5)] is unity in the present example, since each molecule B contains one interaction site. The mass-action expression [eqn. (9)] describing the behaviour of two interacting species at equilibrium, is thus functionally equivalent to the Langmuir equation.

3 Energetics

The following discussion is based on the treatment of solvophobic inter-actions by Sinanoglu,[11] Horvath *et al.*,[12] Melander and Horvath,[13] and Belfort[14] and on the net adsorption energy concept[15] which was, in turn, developed on the basis of the solubility parameter theory.[16]

Non-reactive interactions, excluding those involving magnetic species, are driven mainly by electrostatic forces. The strength of the electric field at the surface will, therefore, determine the strength of interaction between the surface and a sorbate. However, the tendency to undergo sorption is not dictated by the strength of interaction between the sorbate and the sorption site alone. The change in the energy and entropy of the system when sorption takes place is determined by the sum total of displacements which take place when a molecule is transferred from the bulk phase to the surface of the macrospecies. Specifically,

$$G_{sor} = G_{int} - G_{slv} - G_{sl} + G_{ll} \qquad (12)$$

where G is the free energy; sor denotes the overall sorption process; int denotes the direct sorbate–sorbent interaction; slv denotes the solvation of the sorbate, or more precisely the interaction between the sorbate molecule and those solvent molecules which dissociate from it when it attaches to the sorption site; sl denotes the interaction between the sorption site and those solvent molecules which are detached from the surface and replaced by the sorbate in the sorption process; ll denotes the interaction between those solvent molecules which come into contact with each other as a result of the removal of the sorbate molecule from the bulk phase during the sorption process.

Consider the sorption of a molecule from the gas phase. All terms except for G_{int} on the right-hand side of eqn. (12) are negligible. The same holds true for sorption from an apolar solvent. In these instances, therefore, the tendency to sorb will depend primarily on the strength of the interaction between the sorbate and the sorbent. A corollary is that any polar or polarizable sorbate (namely, practically all non-ionic pollutants of interest) will sorb preferentially on surface sites at which a strong electric field exists and the more polar the sorbent the stronger the sorption. When a surface is partially hydrated, some fraction of the polar and ionic sites are occupied by water molecules and adsorption from the gas phase or apolar solvents should be hindered. This was indeed observed. For example, Yaron and Saltzman[17] reported that parathion sorption from hexane on partially hydrated soils decreased as the moisture content increased. The moisture regime may, however, affect sorption in a far more complex manner as is discussed below (Section 6).

When an apolar molecule sorbs from an aqueous solution (or from a solution in any other strongly polar solvent), both G_{int} and G_{slv} are likely to be smaller than G_{ll} and the extent of sorption will be determined by the difference between G_{sl} and G_{ll}. For a given solvent, this difference will be larger the smaller is G_{sl}, namely the less energy is required to remove the polar solvent molecule from the sorption site to make place for the sorbate molecule. This means that sorption of an apolar molecule from a highly polar solvent occurs preferentially at non-polar, or hydrophobic, surface sites. Numerous studies confirm this conclusion.[18,19] The sorption of apolar molecules from water on apolar surfaces is commonly referred to as hydrophobic sorption. Hydrophobic sorption is not the outcome of ill-defined 'hydrophobic forces' but rather, it results from the energy gained by the approach of solvent molecules, previously separated by a solute molecule, to each other after the solute molecule has left the bulk phase to become attached to the surface. The total solvent–solvent interaction energy gained per sorbed molecule will increase with the bulk of that molecule, since more solvent molecules are torn away from each other to allow a bulkier molecule into the liquid phase. Indeed, the solubility of apolar substances decreases and their tendency to sorb from water on hydrophobic sites increases as the apolar molecules become bulkier.[20] Kier and Hall[21] have shown that the first-order molecular connectivity indices, ^1X and ^1X_v, are excellent descriptors for molecular size or bulk. These same descriptors have been found by Nirmalkhandan and Speece[22,23] to correlate well with the aqueous solubility of over 400 organic compounds from a wide variety of chemical classes. These indices have also been cited by several workers[24–26] to correlate well with sorption of non-ionic organics by soils and sediments.

Laird *et al.*,[27] studied the sorption of atrazine on 14 clay samples and concluded that atrazine sorbs from water on smectites mainly as a non-ionic species. The extent of sorption declined as the surface charge density of the smectite increased [namely, as G_{sl}, or more importantly the difference $G_{sl} - G_{int}$ in eqn. (12), increased] in agreement with the above discussion.

'Hydrophobic sorption' which is driven by the energy gained by the removal of apolar species from the aqueous phase is one example of the fact that sorption is not synonymous with surface interaction. Sorption is defined properly as the excess in concentration of a sorbate at a surface over its concentration in the bulk phase, regardless of the cause for this excess.

The reader may wonder why, if actual interactions with the surface are not required for sorption to occur, there should not be an accumulation (*i.e.* 'sorption') of an apolar solute at the air–water or vessel wall–water interface of an aqueous solution. Indeed, such an accumulation does take place. However, the amount of solute 'sorbed' at such interfaces is insignificant when compared to that sorbed on macrospecies in the systems of interest. This is due to the characteristically small surface area of vessels holding aqueous solutions as opposed to the large surface area of environmentally abundant, efficient sorbents (typically hundreds of $m^2\,g^{-1}$). Yet, for solutes of extremely low solubility, hydrophobic sorption to vessel walls is a well-known experimental difficulty.[28]

Bioactive organic monomers are often composed of a hydrophobic backbone and one or more polar moieties. In the unsaturated zone, where both hydrophobic and hydrophilic surface sites exist, such molecules can sorb from an aqueous solution by one of two general mechanisms: (i) through interaction of a polar moiety with polar or charged sites at the sorbent's surface and (ii) hydrophobically, namely, with the hydrophobic backbone of the molecule attached to hydrophobic sites on the macrospecies. If steric factors allow, sorption of a single molecule through both mechanisms (i) and (ii) can take place simultaneously. This is especially true for sorption on organic macrospecies in which both hydrophobic and hydrophilic sites may be found at close proximity (see, for example Sections 4, 5, and 7). While for type (ii) sorption the polar moiety protrudes into the bulk phase and can undergo hydration, sorption of type (i) is a competing process with hydration of the surface and unless simultaneous uptake by both mechanisms takes place, the hydrophobic backbone protrudes into the aqueous solvent. From eqn. (12), it is apparent that mechanism (ii) should dominate the sorption unless the sorbate contains a very highly polar group (polar enough to compete with water for the hydrophilic sorption sites) and the hydrophobic backbone is not too bulky (or simultaneous sorption by both mechanisms is possible). Only if the latter conditions are met, sorption by mechanism (i) may be as (or even more) important as sorption by mechanism (ii). When, on the other hand, the solvent is apolar or sorption occurs in a gas–solid system, sorption by mechanism (i) will always dominate. In the unsaturated zone, the orientation of sorbate molecules at equilibrium relative to the surface may thus depend on whether sorption is from water or if it takes place from a non-aqueous phase. This difference in orientation may, as will be discussed below (see, *e.g.* Section 6), be important to the understanding of the dynamics of small molecules in the unsaturated zone.

4 Organic *vs.* Inorganic Macrospecies as Sorbents

Two broad groups of macrospecies are of great significance as sorbents in the environment, organic and inorganic. While the major inorganic sorbents are characterized by hydrophilic, namely strongly polar, or even charged surfaces, the more important organic macrospecies are rich in both hydrophilic and hydrophobic sorption sites.

Equation (12) enables us to predict the order of sorption of a weakly polar molecule (M) from two solvents, one apolar and the other polar (*e.g.* water) on typical mineral (*e.g.* clay) and organic (*e.g.* a humic acid) surfaces. On the mineral surface, (M) will display more sorption from the apolar solvent that from the polar one. The same molecule is likely to sorb on the organic macrospecies more from the polar than from the apolar solvent due to the high content of hydrophobic sites in the organic sorbent. This expected order of sorption has indeed been demonstrated. For example, Hance[29] studied the sorption of diuron from water and from a petroleum solution on a soil-derived organic material (76% organic matter content) and on an oxidized soil (3% organic matter content). As expected, the sorption of diuron was considerably higher on the oxidized soil when it took place from the apolar solvent. Sorption on the organic sorbent was higher from water. Similarly, Table 1 summarizes the sorption of the polyaromatic hydrocarbon fluorene on two soils and a compost from both water and hexane. While sorption from water correlated well with the sorbents' organic matter content (*i.e.* hydrophobic sites content), sorption from hexane did not correlate with the organic matter content but with the cation-exchange capacity, which is a measure of the charge density of the sorbents. The fundamental difference between sorption of non-ionic molecules from water and from apolar solvents was discussed in detail by Chiou.[30] Due to the similarity between sorption from an apolar solvent and from the gas phase, understanding sorption from apolar solvents has great relevance to the study of the fate of small molecules in the unsaturated zone.

5 Nature of the Interaction Sites

When hydrophobic sorption takes place, the available hydrophobic surface area will define the sorption capacity of the sorbent. When a pesticide is applied to a dry soil and sorption takes place due to interaction between the surface and the sorbate [*i.e.* G_{int} is the dominant term on the right-hand side of eqn. (12)], the higher the electric field strength at the surface the higher the sorption energy. Surface charge density and available hydrophilic surface area will then determine the sorption capacity. Parenthetically, the term 'dry' in the present context does not define a surface which is not hydrated, but rather a surface which is not in contact with a three-dimensional aqueous phase.

The strongest electric field at surfaces is found in the vicinity of ions, if ions are present at the surface. Thus, in the dry state, where G_{int} dominates the sorption process, surface ions are likely to be the most important sorption sites. Exchangeable cations are of particular importance because their very exchangeability implies a greater accessibility than that of non-exchangeable ions. The fact that an exchangeable cation protrudes from the surface, makes it possible for the more polar or polarizable part

Table 1 *Freundlich sorption isotherm parameters for fluorene (from Giat[54])*

Sorbent	Organic matter (%)	Cation-exchange capacity (meq g⁻¹)	Sorption from hexane		Sorption from water	
			K	1/n	K	1/n
Dor soil	1.90	0.61	23.4	0.94	16	0.69
Maagan Michael soil	5.26	0.27	11.9	0.92	66	0.70
Compost	52.70	0.95	143.8	0.86	588	0.90

of the sorbed molecule to approach it, despite possible steric hindrance by other parts of the molecule. A closer approach of the polar or polarizable part of the sorbate molecule to the ion enables a stronger electrostatic interaction.

If a sorbed species contains an anionic group or a group sufficiently polar (or polarizable) to significantly compete with water for sorption at hydrophilic sites, the exchangeable cation will play an important role both in the sorption of this species from water and in its sorption on dry surfaces. The important role exchangeable cations play in sorption was demonstrated by numerous studies.[31,32] Figure 4 summarizes the effect of the exchangeable cation on the sorption of parathion on homionic bentonites from hexane. The strong interaction between exchangeable cations and sorbed molecules, especially in the dry state, may in addition to strengthening the sorption, also induce abiotic degradation of the sorbed molecules.[33]

The role of exchangeable cations in surface interactions of charged minerals such as clays is widely accepted. Less obvious is the role of exchangeable cations in sorption by organic macrospecies. There is a common misconception which views organic matter in the environment as simply hydrophobic. While it is true that organic macrospecies contain most of the hydrophobic sorption sites in the environment, the charge density of the organic matter is rather high (albeit pH dependent). Humic matter in soil, for example, was reported to have a cation-exchange capacity of up to 2.13 meq g^{-1} [34] or more. This cation-exchange capacity is higher than that of the clay minerals (*e.g.* montmorillonite) which contribute most of the cation-exchange capacity in mineral soils. The nature of the exchangeable cation was shown to affect sorption on humic substances,[35] but this effect is complicated by the conformational differences between the various homoionic humic substances.

6 Steric and Temporal Factors

Up to this point, the tendency of a molecule to sorb on a macrospecies at equilibrium was discussed, but equilibrium is rarely reached in the environment. Rate-controlling factors may, therefore, be more important than equilibriuim considerations for predicting the fate and transport of organic chemicals. Steric factors, namely factors associated with the relative position of atoms, play a dominant role in determining the rate of sorption–desorption processes and may also affect the strength and extent of sorption at equilibrium. The role of the protrusion of exchangeable cations in surface interactions[36] was already pointed out above. Following are a few other examples of the importance of steric factors.

At times, extremely small differences in the structure of the interacting species make a considerable difference in the accessibility of sorption sites to sorbate molecules. An instance in point is the retention of substituted phenyls by pillared clays.[37] A 3.3 Å difference in the interlayer spacing of a

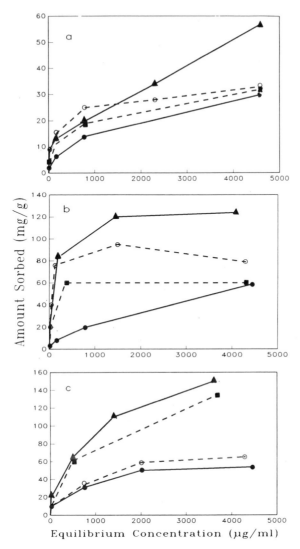

Figure 4 *Isotherms for the adsorption of parathion from* n-*hexane by homionic bentonites at various moisture contents. a. Na–bentonite b. Ca–bentonite c. Al–bentonite. -⊖-, Oven dried (105 °C); —▲—, equilibrated with 15% RH at room temperature (RT); --■--, equilibrated with 32% RH at RT; —●— equilibrated with 47% RH at RT (reproduced from Hayes and Mingelgrin*[35]*)*

cross-linked montmorillonite had a drastic effect on the retention of organophosphate phenyl esters and urons (phenyl substituted ureas) (Table 2). When the retention of a number of disubstituted benzenes on a chromatographic column filled with narrow spaced clays (basal spacing ⩽ 15.3 Å) was measured with hexane as the mobile phase, it was found that the *ortho* isomer was eluted before the *meta* or the *para* isomers.[38]

Table 2 *Retention of some substituted phenyl pesticides on a high-performance liquid chromatographic column. Solid phase: spray-dried
cross-linked Al-montmorillonite; eluent: isopropanol (from Tsvetkov et al.[37])*

Compound	d-*Spacing of solid phase*	
	15.3 Å	18.6 Å
	Capacity ratio (k')*	
Parathion	0.7	4.2
Methyl parathion	0.6	8.1
Paraoxon	0.8	22.3
Linuron	0.1	0.5
Diuron	0.5	2.5
Monuron	2.6	3.6

* The capacity ratio (k') is defined as: $(t_r - t_0)/t_0$, where t_r is the retention time of the
compound of interest, and t_0 is the retention time of non-interacting molecules that are
eluted through the column at the rate of flow of the eluent.

This is in agreeement with the lower penetrability of *ortho* isomers into the
interlayer spaces of smectites and similar solids which was reported by a
number of investigators.[39–41] In the instance of a wide-spaced cross-linked
montmorillonite (basal spacing 18.6 Å) the penetration of the disubstituted
benzenes was practically unhindered. The capacity of some of the *ortho*
isomers to form bidentate complexes with cations strongly enhanced their
sorption on the wide-spaced cross-linked montmorillonite relative to the
sorption of the *meta* and *para* isomers.[37] The relative retention of the
ortho, *meta*, and *para* isomers of disubstituted benzenes on the latter solid
phase was thus dictated directly by the isomers' interaction with the
sorption sites and not by the ease of access to these sites. In contrast, in
the case of the narrow-spaced solid phases, the ease of access did dictate
the retention. Lee *et al.*[42] demonstrated a significant difference between
the penetrability of benzene and toluene (benzene substituted with the
small methyl group) into the interlayers of a Wyoming montmorillonite
saturated with an organic cation (basal spacing 13.8 Å).

Just as small changes in the structure of clays will strongly affect both
the rate and extent of sorbate reaching the interlayer sorption sites, so will
small structural changes in organic polymers affect the accessibility of
sorption sites in the internal cavities of organic coils. Steric factors may
also determine the orientation of the sorbed species relative to the surface.
For example, Theng[43] discussed the sorption of various organic molecules
on clay minerals, pointing out the relationship between the orientation of
the sorbed species and the structure of the sorbent. Structural aspects of
the interaction between clays and organic species were also discussed by
Lagaly.[44]

Environmental conditions will often affect the tertiary structure of macrospecies and thus determine whether sites of limited accessibility in both organic coils and mineral particles are available to sorbate molecules. Coiling–uncoiling of organic macrospecies and swelling or dispersion of clay minerals are controlled by the same forces and are, therefore, affected similarly by environmental factors such as moisture content, pH, and ionic content and strength.[35,45,46]

The effect of the moisture content of bentonite on the equilibrium sorption of parathion from hexane is presented in Figure 4. At low moisture contents, sorption increased with the moisture content due to swelling of the clay which exposed to the sorbate interlayer sorption sites. At higher mosisture contents, a significant fraction of the interlayer sorption sites became blocked (saturated) by the water and sorption decreased.

As the polarity of the solvent increases, it becomes more efficient in competing with the sorbate for sorption sites and in inducing swelling of the sorbent. The combined effect of these two phenomena on the relation between sorbate retention and the polarity of the solvent was demonstrated by Tsvetkov *et al.*[38] The addition of a small quantity of isopropanol to the less polar hexane, increased the retention of substituted benzenes on a Cu-saturated bentonitic solid phase due to the solid phase's swelling (Table 3). A higher portion of isopropanol reduced the retention due to increased competition between isopropanol and the sorbate for the sorption sites.

The effect of pH and the ionic strength and composition on the complexation of atrazine and napropamide with soluble humic acids was described by Lee and Farmer[10] and by Clapp *et al.* (unpublished data). For

Table 3 *Capacity ratios of some substituted benzenes eluted through a Cu–clay column. Mobile phase: (A) hexane, (B) 0.1% isopropanol in hexane, (C) 1% isopropanol in hexane; solid phase: a montmorillonite treated thermally with KCl and then exhanged with Cu (from Tsvetkov* et al.[38]*)*

Compound	Eluent	Capacity ratio*
Nitrobenzene	A	4.5
	B	7.9
	C	1.4
o-Cresol	A	2.4
	B	2.7
	C	0.4
p-Cresol	A	5.7
	B	6.7
	C	0.7
m-Cresol	A	5.5
	B	6.7
	C	0.7

* The capacity ratio (k') is defined in the footnote to Table 2.

the humic acid investigated by Lee and Farmer the complexation of napropamide was hardly affected by the pH. On the other hand, complexation of both atrazine and napropamide by the humic acid investigated by Clapp *et al.* approximately doubled as the pH increased from 4.5 to 7.5. Both studies demonstrated a significant effect of the ionic strength on complexation, the higher the ionic strength the lower was the complexation. Clapp *et al.* reported a decline in complexation of both atrazine and napropamide when the dominant cation in the solution was changed from Na^+ to Ca^{2+} (Table 4). The effect of the above parameters on the tertiary structure (degree of coiling) of the humic macromolecule was apparently a major reason for the observed change in complexation. Both the extent of complexation and the tendency of the organic polymer to uncoil or swell displayed a similar dependence on pH and ionic composition and strength. The complexation of both atrazine and napropamide is likely to be at hydrophobic site on the organic polymer in agreement with the general (albeit not perfect) inverse relation between solubility and complexation with humic acids observed by Lee and Farmer[10] and Clapp *et al.* (unpublished data). A strong direct effect of the pH and ionic parameters on the interaction between the humic acid and the small molecules is, therefore, less probable.

Effect of the Moisture Regime

In the unsaturated zone, wetting and drying cycles are the rule. Two phenomena associated with these cycles may strongly affect the sorption–desorption dynamics of small molecules. One phenomenon is the tendency

Table 4 *Complexation constants of napropamide and atrazine with a water-soluble humic acid in the presence of various cations. Humic acid concentration 2.0 mg ml^{-1}; total napropamide concentration 35 μg ml^{-1}; total atrazine concentration 15 μg ml^{-1}; pH 6.5 (from Clapp, Liu, Zhang, and Mingelgrin, unpublished data)*

Cation	Cation concentration* (mM)	$Kc^\dagger$ (ml mg^{-1})			
		Atrazine		Napropamide	
Na$^+$	10.00	0.14	(0.05)	0.83	(0.04)
	30.00	0.11	(0.02)	0.57	(0.02)
	50.00	0.06	(0.02)	0.44	(0.03)
Ca^{2+}	3.30	0.06	(0.02)	0.05	(0.02)
	10.00	0.03	(0.02)	0.01	(0.00)
	16.70	0.03	(0.02)	0.04	(0.01)
Control		0.12	(0.02)	0.83	(0.02)

* Concentrations of Ca^{2+} and Na$^+$ calculated to give same ionic strengths.
$\dagger$ $Kc = R/C_h$ where R is the ratio between complexed and free monomer and C_h is the total concentration of the humic acid. Values in parentheses are standard deviations.

of the more important sorbents to swell and shrink upon wetting and drying. The other is the difference, discussed above, in the dominant mechanism of sorption between wet and dry systems and the resultant difference in the orientation of the sorbate molecule relative to the surface.

When a system is subject to wetting and drying cycles, it is not always possible for the sorbate molecule to adjust its sorption mode fast enough to follow the fluctuations in moisture content. In the dry state, the more polar part of the molecule will tend to orient itself towards the hydrophilic surface site, leaving the more hydrophobic part of the molecule exposed. Upon rewetting, access of water molecules to the sorption site is hindered by the hydrophobic backbone of the sorbate molecule. This is one possible reason for the slow down in desorption reported by a number of authors[47,48] when small molecules were allowed to incubate in soils for a long time, namely to undergo wetting and drying cycles or at least to be exposed to a drying process. Blockage of polar surface sites to water molecules by the exposed hydrophobic backbone is even more pronounced when polyelectrolytic, large organic molecules interact with the surface. The role of wetting and drying in the formation of water-stable complexes between organic and inorganic soil colloids[49] is most likely a consequence of the establishment of such hydrophobic protection. The rate of wetting or drying, the duration of the wet and dry phases and the nature of the sorbate and sorbent will all determine the rate at which a compound sorbed in one mode of sorption will shift to the other mode as the system fluctuates between wetting and drying.

The exposed hydrophobic backbone is also the reason for the water-repelling property of freshly rewetted, dried humic coils. The many anionic (*e.g.* carboxylate) groups present in these coils interact intramolecularly with di- or poly-valent cations such as calcium, creating upon drying a tight coil with a hydrophobic outer surface. Only after a considerable wetting time do enough water molecules diffuse into the coils to hydrate both cations and anionic groups to an extent that will cause swelling of the coil. The higher the fraction of monovalent counter-cations (excluding protons) in the organic coil, the less tightly it coils upon drying and the higher its tendency to swell or uncoil upon rewetting. The trapping of sorbate molecules during drying in slow-swelling organic coils is another, in many instances dominant, reason for the slow down in desorption upon rewetting.

When sorption takes place in a wet system, the sorbate molecule may diffuse into the inner spaces of an organic coil or mineral aggregate.[50,51] This diffusion is enhanced by the tendency of the macrospecies to swell (or even disperse or uncoil) when wet.[52] If the system is then dried and again rewetted, a considerable apparent hysteresis (reduction in the rate of desorption as compared to that of the original sorption) may occur.[51,53] The reduction in the rate of desorption is more often than not higher than expected from the rate of rewetting (swelling) of the organic coils. One possible explanation for this phenomenon is as follows. The exposed hydrophobic part of molecules sorbed at polar surface sites may create, when it takes place in a bottleneck inside an organic coil or mineral particle, a plug which will considerably slow down rewetting of the internal surfaces (and hence desorption of sorbate molecules) beyond the hydrophobic plug. The very presence of a non-ionic sorbate in the inner spaces of a coil or a particle may thus slow down swelling and with it desorption. Accordingly, Steinberg *et al.*[33] demonstrated the extemely slow release of field-aged ethylene dibromide (EDB) from soils. They assigned the slow release to the entrapment of EDB in soil micropores and demonstrated that the outward effective diffusivity of EDB was at least nine orders of magnitude smaller than expected from adsorption measurements.

The effect of wetting and drying on the sorption–desorption kinetics of slightly soluble molecules is exemplified by the way fluorene sorption and desorption in soil is influenced by the moisture regime of the system.[54] Figure 5 compares the S/C ratios (apparent, or non-equilibrium, sorption coefficients) for fluorene when it was desorbed into water after sorption from hexane on the dry soil, solvent evaporation, and incubation of the fluorene-loaded soil under different moisture regimes. The fluorene-loaded soil underwent one or two wetting and drying cycles, each cycle lasting one week, or was incubated at a constant moisture (air dryness and 120% saturation). The soil (Maagan Michael) was a clay soil, predominantly kaolinitic, with a 5.3% organic matter content. The data presented in Figure 5 indicate that the release of fluorene was considerably slowed down by wetting and drying. When desorption was attempted immediately

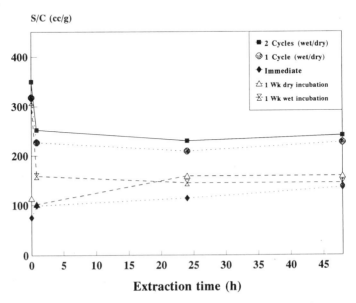

Figure 5 *Desorption kinetics of fluorene from Maagan Michael soil into water after various incubation regimes (reproduced from Giat*[54]*)*

after evaporation of the hexane, the S/C ratio was considerably lower than in the samples which underwent wetting and drying cycles, but the ratio actually increased as shaking with water proceeded. This was interpreted to mean that upon wetting, sites unavailable in the dry state (in which sorption from hexane took place) became accessible to the sorbate due to swelling of organic coils or mineral aggregates. The increased penetration of sorbate molecules into these coils or aggregates resulted in net sorption after the initial release of sorbed fluorene into the water, rather than net desorption.

If penetration of the sorbate into inner surfaces does not reach its equilibrium level by the time drying commences, drying of the system maintains the fraction of sorbate located at the more accessible, outer surfaces above its equilibrium level. Drying in that instance may increase rather than decrease the rate of desorption upon eventual rewetting. How drying actually affects desorption will thus depend on the past history of the system, for example, on the length of time allowed for the initial adsorption from water. Moyer *et al.*[55] reported that three herbicides which were sorbed on peat from water during a 24 hour period displayed a lower rate of desorption after the peat underwent partial drying (equilibrium with 62% relative humidity, achieved in about one month) than when desorption took place directly at the end of the sorption period. For another herbicide the partial drying did not affect desorption significantly. Weber *et al.*[56] reported that more fluoridone desorbed from two soils after dry incubation than after moist incubation. The adsorption period from water was again 24 hours. The three-ringed, relatively bulky fluoridone molecule might have displayed a slow and hence incomplete penetration into intra-particle cavities during the adsorption period. Drying, which brought diffusion into the intra-particle cavities to a halt, caused the fraction of sorbed fluoridone left at the outer surfaces when desorption was eventually attempted to be larger than in the samples incubated moist. In the latter samples, diffusion into intra-particle cavities possibly continued throughout the incubation period.

Desorption of fluorene (initially sorbed from hexane on an air-dried sorbent) into water for 48 hours resulted in higher S/C ratios than those observed when fluorene was sorbed from water for the same length of time[54] although apparent equilibrium was reached in both cases (Table 5). This phenomenon was observed with two soils and a compost. The lower rate of desorption occurred despite the fact that the sorption from hexane was likely to be predominantly on external surfaces, due to the apparently low penetrability of the sorbate into inner spaces in the dry state. Release from external sorption sites was thus a rate-limiting process during desorption when loading was from hexane on a dry sorbent. This suggests that steric hindrance to the exchange between water and the sorbed molecules (caused by the less polar part of the molecule which remains exposed during sorption in the dry state) may affect the kinetics of desorption just as do the dynamics of swelling of the sorbing macrospecies upon wetting.

Table 5 *Comparison between* S/C $(ml\,g^{-1})$ *ratios for sorption and desorption of fluorene in water. S is the amount of fluorene sorbed per unit weight of sorbent and C is the corresponding concentration in solution. Equilibration time was* 48 h *for both sorption and desorption. Desorption carried out after initial sorption from hexane on the dry sorbent and evaporation of the hexane (from Giat[54])*

Sorbent	Sorption	Desorption	t test*
Dor soil	31.26	44.64	+
Maagan Michael soil	112.00	146.89	++
Compost	592.00	806.70	++

* +, ++ Difference significant at 0.05 and 0.01 significance level, respectively.

Even in sorbing systems which are permanently wet, such as sediments, the past history of the system may determine the fate of sorbed molecules if sorption sites of limited access are present.[47] A slow down in the rate of desorption as compared to sorption was observed in some instances[57] in which the system was kept wet throughout. Slow diffusion into inner pores with low accessibility reduced the rate of eventual desorption as compared to the initial sorption. A practical consequence of the above is that if a pollutant which is introduced into a water body diffuses into internal surfaces of a sediment sufficiently slowly, the rate of removal of the pollutant may depend on the time elapsed between the inception of the pollution and cleanup.

The rate of sorption from water on macrospecies which tend to swell and shrink upon wetting and drying, may be strongly dependent on the initial moisture content. When the sorbed molecule is applied in water to an already wet, swollen soil, the inner cavities are already filled with water and sorption into them is diffusion controlled. If the solution is applied to a dry soil, swelling may cause the solute to penetrate the cavities by mass flow.[58] Whether sorption is faster in the initially dry or wet systems depends on whether diffusion or swelling is the faster process. The dependence of the rate of fluorene sorption from water on the initial moisture content of the aformentioned Maagan Michael soil[54] is presented in Figure 6. Sorption was faster on the initially dry soil, except in the first few minutes of contact. Swelling of the macrospecies seems to be more rapid than diffusion of the sorbate into the inner surfaces in this particular soil.

Both sorption and desorption rates strongly depend on the past moisture regime of the system. It is apparent that the transport and availability of pollutants in the unsaturated zone are affected by exposure to wetting and drying cycles, the moisure content and at the time the pollutant was introduced and the time which elapsed between sorption and desorption. Predictive transport models based on the assumption of instantaneous sorption–desorption equilibrium are thus likely to fail. Even those models

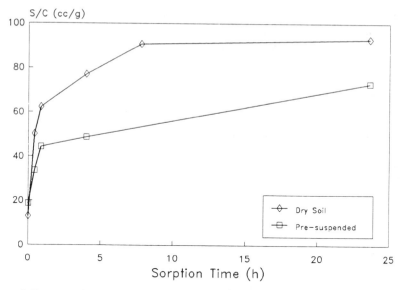

Figure 6 *Sorption kinetics of fluorene from water on Maagan Michael soil as a function of the soil's initial moisture content (reproduced from Giat[54])*

which describe the time dependence of the sorption–desorption process by a simple low-order rate equation are bound to fail in many instances; for example, whenever the unsaturated zone is rich in mineral or organic sorption sites of limited and variable access.

7 The Partition Hypothesis, K_{oc}, and Related Concepts

The uptake from water of non-ionic substances by the solid fraction of soils, sediments, and other environmental compartments has been attributed to a partition between the aqueous phase and a three-dimensional hydrophobic phase present in the solid fraction.[5,59–63] The envisioned partition process is something akin to polymer diffusion. Indeed, many observations related to sorption of non-ionic compounds were successfully explained or even predicted with the aid of the partition theory.

It was argued above that sparingly soluble molecules will tend to sorb from water at hydrophobic sites. Whenever a three-dimensional hydrophobic network is present, such molecules will indeed partition into it. However, in most soils and sediments, the existence of a significant amount of a three-dimensional hydrophobic network is yet to be demonstrated. Such a network does exist in systems sometimes used to mimic uptake by soils or sediments, such as the C18 HPLC solid phase.[64] Furthermore, the same considerations which dictate partition into a three-dimensional hydrophobic phase will also dictate sorption at the interface between a hydrophobic surface site and water. It is not very easy to distinguish experimentally between the two-dimensional sorption and the three-dimensional partition processes.

A number of critical evaluations of the partition hypothesis have been already presented.[65-68] To these arguments it will suffice to add the following. The polyelectrolytic nature of the organic sorbents which dominate the uptake of non-ionic species in soils is well accepted.[45] Namely, hydrophobic sites exit along the macromolecular string intermingled with polar or ionic groups (as opposed to the uniform apolar nature of say, C18). As discussed above, the swelling and shrinking (or uncoiling and coiling) of the organic macrospecies in response to changes in environmental parameters and the related performance of these macrospecies as a sorbent are derived to a large extent from their polyelectrolytic nature. The hysteresis (or apparent hysteresis) often observed[47] in the sorption–desorption process is in many instances the outcome of the polyelectrolytic nature of the organic fraction. The role of wetting and drying in retarding desorption from sorbate-loaded expandable or swelling macrospecies has been discussed in detail above. Hysteresis in the sorption–desorption process and in particular the hysteresis observed after drying and rewetting, cannot be easily reconciled with the partition hypothesis.

It is hard to view the complexation of non-ionic species with the relatively small soluble macromolecules as a partition process. As discussed above, this compelxation is, as a rule, due to 'hydrophobic sorption'. There is no reason why a sorption process similar to that occurring on soluble macromolecules should not occur on larger macrospecies (such as insoluble humic coils) which contain hydrophobic sites.

Explaining the saturation observed in the complexation isotherm of apolar molecules with soluble humic acids (Figures 2 and 3) as well as in the sorption isotherms of apolar molecules on soils (see Weber and Miller[9] for an extensive compilation of examples) without invoking in all those instances a strong deviation from ideality is a challenge to supporters of the partition theory.

One practical conclusion which evolved from the partition hypothesis is the constancy of the K_{oc} parameter. It was argued that since uptake of apolar molecules in soils or sediments is predominantly by the sorbent's organic fraction, and since the uptake process is a partition into a three-dimensional hydrophobic network rather than an interaction with specific sites, the tendency of a molecule to partition should simply depend on the organic matter content. The parameter K_{oc} was thus defined

$$K_{oc} = K_d/f_{oc} = S_{eq}/(C_{eq}f_{oc}) \qquad (13)$$

where K_d is the sorption constant, C_{eq} is the equilibrium concentration of the sorbate in the aqueous phase, S_{eq} is the amount sorbed per unit weight of the solid phase at equilibrium, and f_{oc} is the fraction of organic carbon in the soild phase. It is often assumed[5,62,69] that K_{oc} is, to a good approximation, a constant for a given non-ionic sorbate. Once K_{oc} is measured, sorption in any soil or sediment (K_d) can be obtained from the product of K_{oc} and the sorbent's organic carbon content.

Unfortunately, the assumption of constancy of K_{oc} failed in many instances (*e.g.* Table 6). Furthermore, an inverse relation was observed, especially in low organic matter soils between f_{oc} and the observed K_{oc}.[18,70,71] This is at least in part due to the contribution of the mineral fraction of the soil to sorption. On the other hand, the approximate constancy of K_{oc} is likely to be maintained within a family of soils or sediments close enough in properties, indicating that K_{oc} can be a useful empirical parameter if it is used properly.

Even within a set of sorbing media for which K_{oc} is a constant, the transport and availability of organic molecules cannot always be predicted on the basis of K_{oc} alone. When sorption is studied under realistic conditions rather than by such procedures as equilibrium batch measurements commonly used to derive K_{oc}, the inadequacy of this parameter becomes obvious.[47]

Table 6 K_{om} *Values of non-ionic compounds in various soils.* K_{om} = $K_{oc}/1.72$ *(from Mingelgrin and Gerstl*[67]*)*

Compound	K_{om} (ml g^{-1} OM)	Organic matter (%)
EDB	21–93[a,b]	0.5–21.7
Piperophos	72–7,627[c]	1.8–10.2
Napropamide	110–1,223[b,d]	0.1–2.4
Parathion	182–9,200[e–g]	0.2–6.1
Phorate	211–3,980[f,h]	0.2–31.7
Lindane	427–1,502[i–k]	1.2–20.5
Disulfoton	470–5060[f,l,m]	0.2–4.6
Chlorpyrifos	1,255–20,400[h,j]	1.2–6.6
DDT	76,300–257,040[j,n,o]	1.6–3.9
3-Methycholanthrene	211,000–3,710,000[p]	0.8–4.1[q]
Dibenzanthracene	328,500–1,779,000[p]	0.8–4.1[q]

[a] J.W. Hamaker and J.M. Thompson, 'Organic Chemicals in the Soil Environment' eds. C.A.I. Goring and J.W. Hamaker, Marcel Dekker, New York, 1972, p. 49.
[b] U. Mingelgrin and Z. Gerstl, *J. Environ. Qual.*, 1983, **12**, 1.
[c] U. Hata and Y. Isozaki, *J. Pest. Sci.*, 1980, **5**, 23.
[d] C.H. Wu, N. Buehring, J.M. Davidson and P.W. Santelmann, *Weed Sci.*, 1975, **23**, 454.
[e] J.W. Biggar, U. Mingelgrin and M.W. Cheung, *J. Agric. Food Chem.*, 1978, **26**, 1306.
[f] P.H. King and P.L. McCarty, *Soil Sci.*, 1968, **106**, 248.
[g] B. Yaron and S. Saltman, *Soil Sci. Soc. Am. Proc.*, 1972, **36**, 583.
[h] P. Felsot and P.A. Dahm, *J. Agric. Food Chem.*, 1979, **27**, 557.
[i] B.D. Kay and D.E. Elrick, *Soil Sci.*, 1967, **104**, 314.
[j] P.J. McCall, R.L. Swann, D.A. Laskowski, S.M. Unger, S.A. Vrona and H.J. Dishburger, *Bull. Environ. Contam. Toxicol.*, 1980, **24**, 190.
[k] A.C. Mills and J.W. Biggar, *Soil Sci. Soc. Am. Proc.*, 1969, **33**, 210.
[l] I.J. Graham-Bryce, *Soc. Chem. Ind. Monogr.*, 1968, **29**, 251.
[m] I.J. Graham-Bryce, *J. Sci. Food Agric.*, 1969, **20**, 489.
[n] C.T. Chiou, L.J. Peters and V.H. Freed, *Science*, 1979, **206**, 831.
[o] Y. Shin, J.J. Chodan and A.R. Waleott, *J. Agric. Food Chem.*, 1970, **18**, 1129.
[p] J.J. Hassett, J.C. Means, W.L. Banwart and S.G. Wood, EPA Report, Sorption Properties of Sediments and Energy-Related Pollutants, 1980.
[q] Lowest organic matter samples not included.

8 Limitations of the Generalized Approach

The various sorbents in the environment vary widely in their chemical and physical properties. Yet, the sorption process follows the same physical rules regardless of the sorbent. Expandable clays and organic macrospecies are probably the two most important classes of sorbents in soils and sediments. Swelling and dispersion of clay tactoids and swelling and uncoiling of humic soils are affected by factors such as moisture content, pH, and ionic composition and strength in a similar manner. Drying makes both the interlayers of clays and the inner cavities of organic coils less accessible due to shrinkage. Similarly, the higher the concentration of di- or poly-valent cations, the lower the tendency of both expandable clays and humic acids to swell when wetted.[35]

Analogies should not, however, be carried too far. The strong ionic contribution to the bonds in the crystals of most minerals which are important sorbents in the environment, makes the mineral surfaces mainly hydrophilic. The commonly occurring phenomenon of isomorphous subtitution enhances this hydrophilicity. Humic substances, however, are rich in both hydrophilic and hydrophobic sorption sites. Organic (*e.g.* humic) substances may respond to basic conditions by hydrolytic transformations and dissolution. Clays are considerably less sensitive to basic conditions, although clay hydrolysis and dispersion is also a pH-dependent process.

Although the tendency of a molecule to sorb at a given site is independent of the size of the sorbing macrospecies, this size may strongly affect the rate and at times the extent of sorption. As a rule, the larger the macrospecies the smaller the available surface area per unit weight. The fraction of sorption sites with a restricted accessibility should increase with the size of the macrospecies. Steric hindrance is likely, therefore, to play a lesser role in complexation with highly soluble humic macromolecules than in sorption on the, usually larger, insoluble humic macrospecies.

The shape of the macrospecies should also make a difference in its sorptive behaviour. For example, the accessibility of a higher fraction of sorption sites should be more controlled by environmental parameters for stringy humic macrospecies than for the laminar, swelling clays due to the capacity of the former to coil and uncoil. Environmental parameters should have only a small effect on the accessibility of sorption sites for those mineral or mixed mineral–organic aggregates which are devoid of internal sorption sites having variable accessibilities.

9 Conclusions

A few simple rules dictate the equilibrium sorption of small molecules regardless of the nature of the sorbent. Yet, kinetic and steric factors make sorption in the environment hard to predict quantitatively. As far as transport, availability, and susceptibility to biotic degradation is concerned, the dynamics of a pollutant's sorption and desorption is likely to be more important than its sorption at equilibrium. The complicated dependence of both the tertiary structure of the major sorbents in the environment and the sorption process itself on environmental parameters makes predictive modelling even more complex.

References

1. D.E. Kile and C.T. Chiou, 'Aquatic Humic Substances', eds. I.H. Suffet and P. MacCarthy, American Chemical Society, Washington, 1989, p.130.
2. J.A. Vinten, B. Yaron and P.H. Nye, *J. Agric. Food. Chem.*, 1983, **31**, 662.
3. R.L. Wershaw, P.J. Brucar and M.C. Goldberg, *Environ. Sci. Technol.*, 1969, **3**, 271.
4. C.T. Chiou, R.L. Malcolm, T.I. Brinton and D.E. Kile., *Environ. Sci. Technol.*, 1986, **20**, 502.
5. P.M. Gschwend and S. Wu, *Environ. Sci. Technol.*, 1985, **19**, 90.
6. D.C. Bouchard, C.G. Enfield and M.D. Piwoni, 'Reactions and Movement of Organic Chemicals in Soils', eds. B.L. Sawhney and K. Brown, Soil Science Society of America, Madison, Wisconsin, 1989, p.349.
7. S. Brunauer, P.H. Emmett and E. Teller, *J. Am. Chem. Soc.* 1938, **60**, 309.
8. U. Mingelgrin and F. Tsvetkov, *Clays Clay Min.*, 1985, **33**, 62.
9. J.B. Weber and C.T. Miller, 'Reactions and Movement of Organic Chemicals in Soils', eds. B.L. Sawhney and K. Brown, Soil Science Society of America, Madison, Wisconsin, 1989, p. 305.
10. D.U. Lee and W.J. Farmer, *J. Environ. Qual.*, 1989, **18**, 468.
11. O. Sinanoglu, 'Molecular Associations in Biology', ed. B. Pullman, Academic Press, New York, 1968, p. 427.
12. C. Horvath, W. Melander and I. Molnar, *J. Chromatogr.*, 1976, **125**, 129.
13. W. Melander and C. Horvath, 'Activated Carbon Adsorption of Organics from the Aqueous Phase', eds. I.H. Suffet and M.J. McGuire, Ann Arbor Science Publishers, Ann Arbor, MI, 1980, Vol. 1, p. 65.
14. G. Belfort, 'Chemistry in Water Reuse', ed. W.J. Cooper, Ann Arbor Science Publishers, Ann Arbor, MI, 1981, Vol. 2, p. 207.
15. M.J. McGuire and I.H. Suffet, 'Activated Carbon Adsorption of Organics from the Aqueous Phase', eds. I.H. Suffet and M.J. McGuire, Ann Arbor Science Publishers, Ann Arbor, MI, 1980, Vol. 1, p. 91.
16. J.H. Hildelerand and R.L. Scott, 'The Solubility of Non-Electrolytes', Reinhold, New York, 1950.
17. B. Yaron and S. Saltzman, *Soil Sci. Soc. Am. Proc.*, 1972, **36**, 583.
18. J.J. Hassett and W.L. Banwart, 'Reactions and Movement of Organic Chemicals in Soils', eds. B.L. Sawhney and K. Brown, Soil Science Society of America, Madison, Wisconsin, 1989, p. 31.
19. K.B. Woodburn, P.S.C. Fao, M. Fukui and P. Nkedi-Kizza, *J. Contam. Hydrol.*, 1986, **1**, 227.
20. D. MacKay, *Environ. Sci. Technol.*, 1977, **11**, 1219.
21. L.B. Kier and L.H. Hall, 'Molecular Connectivity in Structure-Activity Analysis', Research Studies Press, Letchworth, 1986.
22. N.N. Nirmalakhandan and R.E. Speece, *Environ. Sci. Technol.*, 1988, **22**, 328.
23. N.N. Nirmalakhandan and R.E. Speece, *Environ. Sci. Technol.*, 1989, **23**, 708.

24. D.A. Bahnick and W.J. Doucette, *Chemosphere*, 1988, **17**, 1703.
25. Z. Gerstl, *J. Contam. Hydrol.*, 1990, **6**, 357.
26. A. Sabljic, *J. Agric. Food Chem.*, 1984, **32**, 243.
27. D.A. Laird, E. Barriuso, R.H. Dowdy and W.C. Koskinen, *Soil Sci. Soc. Am. J.*, 1992, **56**, 62.
28. L.W. Lion, T.B. Stauffer and W.G. MacIntyre, *J. Contam. Hydrol.*, 1990, **5**, 215 and references therein.
29. R.J. Hance, *Weed Res.*, 1965, **5**, 108.
30. C.T. Chiou, 'Toxic Organic Chemicals in Porous Media', eds. Z. Gerstl, Y. Chen, U. Mingelgrin and B. Yaron, Springer Verlag, Berlin, 1989, p. 163.
31. R.L. Praffit and M.M. Mortland, *Soil Sci. Soc. Am. J.*, 1968, **32**, 355.
32. B.T. Bowman and W.W. Sans, *Soil Sci. Soc. Am. J.*, 1977, **41**, 514.
33. N.L. Wolfe, U. Mingelgrin and G.C. Miller, 'Pesticides in the Soil Environment', ed. H.H. Cheng, Soil Science Society of America, Madison, Wisconsin, 1990, p.103.
34. H.L. Bohn, B.L. McNeal and G.A. O'Connor, 'Soil Chemistry', Wiley, New York, 1979.
35. M.H.B. Hayes and U. Mingelgrin, 'Interactions at the Soil Colloid–Soil Solution Interface', eds. G.H. Bolt, M.F. Boodt, M.H.B. Hayes and M.B. McBride, Kluwer Academic Publishers, Dordrecht, 1991, p. 323.
36. U. Mingelgrin, S. Zaltzman and B. Yaron, *Soil Sci. Soc. Am. J.*, 1977, **41**, 519.
37. F. Tsvetkov, U. Mingelgrin and N. Lahav, *Clays Clay Miner.*, 1990, **38**, 380.
38. F. Tsvetkov, L. Heller-Kallai and U. Mingelgrin, *Clays Clay Miner.*, 1993, **41** in the press.
39. S. Namba, Y. Kanai, H. Shoji and T. Yashima, *Zeolites*, 1984, **4**, 77.
40. R.S. Richards and L.V.C. Rees, *Zeolites*, 1988, **8**, 35.
41. D. Sybliska and E. Smolkova-Keulemansova, 'Application of Inclusion Compounds in Chromatography', eds. J.L. Atwood, J.E.D. Davies and D.D. MacNicol, Academic Press, London, 1984, p.173.
42. J.F. Lee, M.M. Mortland, C.T. Chiou, D.E. Kile and S.A. Boyd, *Clays Clay Miner.*, 1990, **38**, 113.
43. B.K.G. Theng, 'The Chemistry of Clay–Organic Reactions' Adam Hilger, London, 1974.
44. G. Lagaly, Proceedings of the International Clay Conference, 1985, eds. L.G. Schultz, H. van Olphen and F.A. Mumpton, The Clay Mineral Society, Bloomington, Indiana, 1987, p.343.
45. M.H.B. Hayes and G.H. Bolt, 'Interactions at the Soil Colloid–Soil Solution Interface' eds. G.H. Bolt, M.F. De Boodt, M.H.B. Hayes and M.B. McBride, Kluwer Academic Publishers, Dordrecht, 1991, Chap. 1, p. 1.
46. R.G. Gast, 'Mineral in Soil Environments', eds. J.B. Dixon and S.B. Weed, Soil Science Society of America, Madison, Wisconsin, 1977, p.27.
47. J.J. Pignatello, 'Reactions and Movement of Organic Chemicals in Soils', eds. B.L. Sawhney and K. Brown, Soil Science of America, Madison, Wisconsin, 1989, p.45.
48. S.L. Scribner, T.R. Benzing, S. Sun and S.A. Boyd, *J. Environ. Qual.*, 1992, **21**, 115.
49. M.H.B. Hayes and F.L. Himes, 'Interactions of Soil Minerals with Natural Organics and Microbes', eds. P.M. Huang and M. Schnitzer, Soil Science Society of America, Madison, Wisconsin, 1986, p. 103.
50. J.A. Leenheer and J.L. Ahlrichs, *Soil Sci. Soc. Am. Proc.*, 1971, **35**, 700.

51. D. Mackay and B. Powers, *Chemosphere*, 1987, **16**, 745.
52. W.G. Lyon and D.E. Rhodes, 'The Swelling Properties of Soil Organic Matter and their Relation to Sorption of Non-Ionic Organic Compounds', Project-report EPA-600/2-91/003, Ada, Oklahoma, 1991.
53. S.M. Steinberg, J.J. Pignatello and B.L. Sawhney, *Environ. Sci. Technol.*, 1987, **21**, 1201.
54. V. Giat, 'The Interaction of Organic Molecules with the Solid Phase as Affected by Soil Moisture Regime', Hebrew University of Jerusalem, Thesis submitted for the degree 'Master of Science', 1992.
55. J.R. Moyer, R.B. McKercher and R.J. Hance, *Can. J. Soil Sci.*, 1972, **52**, 439.
56. J.B. Weber, P.H. Shea and S.B. Weed, *Soil Sci. Soc. Am. J.*, 1986, **50**, 582.
57. S.A. Clay and W.C. Koskinen, *Weed Sci.*, 1990, **38**, 74.
58. E.G. Young and P.B. Leeds-Harison, *J. Soil Sci.*, 1990, **41**, 665.
59. C.T. Chiou, L.J. Peters and V.H. Freed, *Science*, 1979, **206**, 831.
60. C.T. Chiou, P.E. Porter and D.W. Schmedding, *Environ. Sci. Technol.*, 1983, **17**, 227.
61. C.T. Chiou, T.D. Shoup and P.E. Porter, *Org. Geochem.*, 1985, **8**, 9.
62. C.T. Chiou, 'Reactions and Movement of Organic Chemicals in Soils', eds. B.L. Sawhney and K. Brown, Soil Science Society of America, Madison, Wisconsin, 1989, p. 1.
63. S.W. Karickhoff, *J. Agric. Food Chem.*, 1981, **29**, 425.
64. J.G. Dorsey and D.A. Dill, *Chem. Rev.*, 1989, **89**, 331.
65. S. Banerjee, S.H. Yalkowsky and S.C. Valvani, *Environ. Sci. Technol.*, 1980, **14**, 1227.
66. B.G. Kyle, *Science*, 1981, **213**, 683.
67. U. Mingelgrin and Z. Gerstl, *J. Envrion. Qual.*, 1983, **12**, 1.
68. P.A. Wahid and N. Sethunathan, *J. Agric. Food Chem.*, 1981, **29**, 425.
69. R.E. Green and S.W. Karickhoff, 'Pesticides in the Soil Environment: Processes, Impacts and Modelling', ed. H.H. Cheng. Soil Science Society of America, Madison, Wisconsin, 1990, Chap. 4, p. 79.
70. Z. Gerstl and U. Mingelgrin, *J. Environ. Sci. Health*, 1984, **B19**, 297.
71. C. Means, G. Wood, J. Hassett and W.L. Banwart, *Environ. Sci. Technol.*, 1982, **16**, 93.

6

Recent Advances in Sorption Kinetics

By Joseph J. Pignatello

THE CONNECTICUT AGRICULTURAL EXPERIMENT STATION, DEPARTMENT OF
SOIL AND WATER, NEW HAVEN, CONNECTICUT 06504-1106, USA

1 Introduction

Sorption is fundamental to the fate of chemicals in the terrestrial environment. Despite intensive research, we continue to uncover its complexities. One of these clearly is kinetics. It has become apparent over the past decade or so that sorption time scales in soils, aquatic sediments, and aquifer materials are often greater than several days, and can be as long as years. Consequently, sorption/desorption may be rate-limiting to dissipation mechanisms.

This phenomenon has been variously called 'kinetic sorption', 'resistant sorption', 'non-equilibrium sorption', and 'slowly reversible sorption'. Obviously, its rate-limiting nature has not always been widely recognized. Of the perhaps tens of thousands of experiments, nearly all have been, and are still, carried out with equilibration times of 72 h or less, most under 24 h. The nomenclature itself reflects a bias: the term 'non-equilibrium' conveys that equilibrium is the normal and expected condition; modellers speak of testing the LEA ('local equilibrium assumption'). Partly responsible for the confusion is the two-stage nature of sorption in most systems, *i.e.* a fast step followed by a slow one.

The subject of slow sorption kinetics was reviewed a few years ago by me[1] and others.[2] After a brief early history, I will attempt to describe, using recent literature wherever possible, the implications that slow sorption has for contaminant transport, bioavailability, long-term persistence, remediation, and analytical methodology. Next I will discuss advances in mechanistic understanding. Finally, I will discuss research needs.

Before the mid 1980s there was little discussion of sorption kinetics of organic compounds in geological materials. A few noteworthy citations prior to this time are given. One of the first comments to appear was a hypothesis in 1965,[3] unaccompanied by data, that 'steric shielding' might account for the apparent bio-recalcitrance of some organic compounds in soils. A few years later a review of sorption[4] stated that '. . . there have apparently been no studies of the kinetics of adsorption in soil . . .' apart from one study using pure clays. In 1977 Cameron and Klute[5] published a seminal paper advancing what was later called the 'two-site' model for transport of sorbing chemicals in a soil column. This model suggested a combination of equilibrium and first-order kinetic terms to explain break-through curve asymmetry, or 'tailing' observed in elution curves.

In the early 1980s, Karickhoff and co-workers published a series of papers[6-9] describing the two-stage fast–slow sorption and desorption of hydrophobic polyaromatic hydrocarbons (PAHs) and chlorinated benzenes in sediment suspensions. Karickhoff stated in a 1984 review article on sorption thermodynamics[7] that: '. . . this author is very perplexed by the failure of most studies to acknowledge any evidence of 'resistant' sorption . . .' A few other papers appeared around that time describing similar behaviour (citations in reference 1). Finally, a well-cited paper published in 1986[10] advanced a micropore diffusion model to describe chlorinated aromatic uptake by sediment particles. Since the mid 1980s many studies have been published covering field observations, mathematical modelling, and mechanism; slow sorption now appears to be recognized as a major complicating factor in predicting fate and transport behaviour and in remediation of contaminated sites. As one study recently pointed out,[11] dissipation of pesticides in soil generally fails to follow simple first-order kinetics.

2 Implications for Fate and Transport

Transport

The movement of dissolved chemicals through a porous soil can be described by the advection-dispersion-reaction (ADR) equation, which in one dimension is:

$$\delta C/\delta t + \beta/\theta \, (\delta S/\delta t) = D_h \, (\delta^2 C/\delta z^2) - v(\delta C/\delta z) + \delta f(C, S)/\delta t \quad (1)$$
$$\text{(dispersion)} \qquad \text{(advection)} \quad \text{(reaction)}$$

where C and S are the aqueous and sorbed concentrations, t is time, z is distance, β is bulk density, θ is porosity, D_h is the hydrodynamic dispersion coefficient, v is flow velocity, and $f(C, S)$ is some function of the concentrations. The right-hand terms describe, respectively, hydro-dynamic dispersion, advection, and reaction or other loss mechanism. This equation applies to movement in the dissolved state, but analogous equations apply to vapour transport.

The sorbed concentration S appears in the second term on the left and in the reaction term. Ignoring reaction for the moment, eqn. (1) cannot be solved without knowing how S changes with time. This is the essential problem. If sorption is fast, the $\delta S/\delta t$ term is simply replaced by a Freundlich or other equilibrium expression that includes a sorption parti-tion coefficient (K_p), and can be folded into the C term.

Several studies have shown how unrealistic it can be to assume equilibrium in a 24 or even 72 h period. For example, the apparent sorption constant, K_{app}, of residual atrazine and metolachlor in field-treated agricultural soils collected a few months after application was up to 42 times greater than K_p from 24 h sorption isotherms.[12] Values of K_{app} increased with time in the environment. Analogous non-equilibrium behaviour was observed for field-weathered simazines;[13] a few weeks after application, only about 5% of the simazine predicted by the 24 h K_p was present in soil water. Other examples of serious deviations from 'equilibrium' behaviour for field-contaminated soils are reported by Pavlostathis and Mathavan[14] for trichloroethene (TCE), tetrachloroethene (PCE), toluene, and xylene; and by Smith *et al.*[15] for TCE. Additional examples were cited earlier.[1]

Other environmental media in which slow sorption occurs include aquifer sediments, biomass, and dissolved organic matter (DOM). Incubation of low organic carbon (OC) aquifer sediments (0.19% OC) with PCE and 1,2-dibromo-3-chloropropane (DBCP) for 6 or 30 d at *ca.* 20 mg l^{-1} left a small residual fraction that did not desorb for at least 35 d.[16] Ball and Roberts[17,18] studied long-term sorption of halogenated hydrocarbons, PCE, and 1,2,4,5-tetrachlorobenzene (TeCB), in aquifer sediments containing 0.02% OC. Depending on the particle size fraction, PCE required 10 or more days and TeCB required 100 or more days to equilibrate. The ultimate K_p was determined from pulverized material, which equilibrated faster. Chitin, an amino–sugar polymer abundant in nature, sorbed lindane hysteretically.[19] Even desorption from DOM is slower than expected. The desorption rate constant of tetrachlorobiphenyl from DOM (Aldrich humic acid) was found to be 0.004 min^{-1} ($t_{1/2} = 200$ min).[20] Sorption of napropamide [2-(α-naphthoxy)-N,N-diethylpropionamide] on dissolved peat humic acid was 'not entirely reversible' after 4 d.[21]

Field evidence for the effects of sorption kinetics on transport exists. Slow sorption may have contributed to the transport behaviour of deliberately injected halogenated hydrocarbons in an aquifer in Canada.[22] In an important observation, it was found that the organic plumes slowed down with time in the aquifer, compared to the bromide and chloride tracer plumes. These investigators have suggested slow sorption as a cause.[23] Indeed, the materials referred to above in the PCE and TeCB study[17,18] were from the same aquifer.

Obviously, the neglect of sorption kinetics can lead to tremendous errors in predicting contaminant movement. It is typically found that 'two-compartment' uptake or transport models for sorbing chemicals work best. The most frequently used model is the 'two-site' mode, in which S_1 is in rapid-reversible equilibrium, while S_2 exchanges with first-order kinetics [eqn. (2)].

$$C \underset{K_1}{\rightleftharpoons} S_1 \underset{k_{-2}}{\overset{K_2,\,k_2}{\rightleftharpoons}} S_2 \tag{2}$$

with equilbrium expressions:

$$S_1 = K_1 C; \quad S_2 = K_2 S_1; \quad S_1 + S_2 = K_{eq} C \tag{3}$$

and rate law:

$$\delta S_2/\delta t = k_2 S_1 - k_{-2} S_2 \tag{4}$$

where K_1, K_2, and K_{eq} are equilibrium constants for the equilibrium domain, the kinetic domain, and total sorbate, respectively; and k_2 and k_{-2} are first-order rate constants. It makes no difference mathematically whether exchange with S_2 occurs with S_{max} directly with C. Equation (4) is the rate law used to solve the ADR equation [eqn. (1)].

An analogous two-compartment equilibrium–diffusion model has been used,[24] where now the rate constant is an effective diffusion coefficient, D_{eff}:

$$\begin{array}{cc} & K_2 \\ K_1 & D_{eff} \\ C \rightleftharpoons S_1 & \leftrightarrow S_2 \end{array} \tag{5}$$

The equilibrium expressions are the same as the two-site model in eqn. (3). Imagine the column to be composed of uniform sorbent spheres of radius, a. The rate of change of s_D, the local sorbate concentration per unit volume sorbent 'particle' at distance r from the centre, is given by:

$$\delta s_D/\delta t = D_{eff}/r \, [\delta^2(r s_D)/\delta r^2] \tag{6}$$

Assuming no change in s_D at $r = 0$, and equilibrium at the sorbent–liquid interface, the volume-averaged concentration in the diffusion domain is:

$$S_2(t) = 3\theta/(\beta a^3) \int_{r=0}^{a} s_D(t, r) \, r^2 \delta r \tag{7}$$

Sorbate in the fast state at equilibrium may comprise only a few percent of total sorbed material: 0–30% in one study[18] and 8–18% in another,[24] assuming true equilibrium had been achieved. The point here is that the experimental K_p from a short-term sorption experiment may reflect sorption in the equilibrium domain instead of sorption as a whole (*i.e.* K_1 rather than K_{eq}). For example,[24] upon fitting the elution curves of two herbicides to the diffusion model [eqn. (5)], K_1 came to be within a factor of 1.25 of the 24 h K_p, but K_{eq} was estimated to be 5–10 times greater than the 24 h K_p.

Bioavailability

The sink term in eqn. (1) indicates that the reaction rate of molecules in the sorbed state must be known. Several studies have shown that micro-organisms are much less able to take up sorbed than dissolved molecules.[25-27] Not surprisingly then, biodegradation rates are strongly affected by sorption kinetics. Some time ago, we showed that aged field residues of ethylene dibromide (EDB) were practically inert compared to added ^{14}C-EDB toward biodegradation.[28] Scribner *et al.*[13] found that 48% of added simazine was biodegraded in 34 d, whereas no biodegradation of native simazine at about the same concentration occurred.

Availability for uptake by higher organisms is likewise an issue here. Half of all sugar beet seedlings tested were damaged by a 0.22 mg kg^{-1} spike of simazine to the soil, but field residues at the same concentration had no effect.[13] Freshly added tritium-labelled benzo[*a*]pyrene (^{3}H-BaP) and its metabolites in a sediment were far more bioavailable to both clams and amphipods than solvent-extractable native BaP and metabolites.[29]

Long-term Persistence

The persistence of EDB at trace levels in certain soils for as long as 20 years beyond the last fumigation of the soil[30] attests to the extremely slow desorption rates that are possible, considering that EDB is volatile, weakly sorbing, fairly water soluble, and biodegradable under aerobic conditions. Wild *et al.*[31] analysed PAH in archived agricultural soil samples which had received sewage sludge ammendments from 1942 to 1961. The fall-off in concentration after 1961 compared to the control indicated a mean half-life for the sum of PAH of 19 years, with a range of 8–28 years for individual compounds. These values were much greater than the ~2 year half-lives reported for spiked samples. They hypothesized that following each sludge addition there was relatively rapid loss of PAHs, ultimately leaving a more slowly degraded fraction. A similar two-stage biodegradation of surfactants in sludge-treated soil was observed, where the slow stage lasted beyond the experimental time frame of about 350 d.[32]

Remediation

Slow sorption has come to be recognized as a serious impediment to bio or conventional remediation of contaminated soils and aquifers. In a feature article on pump-and-treat methods for remediating groundwater, MacKay and Cherry[33] conclude that this common approach is likely to require exceedingly long times. Part of their rationale was that:

'...kinetic limitations to desorption can occur as observed in field studies ..., thereby increasing both the time to achieve clean-up and the total volume of water that must be extracted to flush the contaminated zone. Furthermore, if pumping is ceased before all of the contaminant is removed, the contaminant concentration in the groundwater will rise as desorption continues.'

A recent Workshop on Bioremediation Needs[34] recognized slow sorption as a major problem for soils clean-up:

'Many ... pollutants that are quickly degraded under laboratory conditions persist in the environment ... This constraint on bioremediation represents a major limitation to widespread use of many biotechnologies.'

The report gave high priority to research on the role of sorption kinetics in biodegradation rates in soils.

Analytical Methodology

The ability to efficiently extract contaminants from environmental matrices is obviously crucial to understanding and predicting their behaviour. The most common method is Soxhlet extraction in which the sample placed in a porous thimble is repeatedly batch-leached with condensed solvent vapours. Simple room-temperature batch extraction or sonication-extraction is also often used. These methods are validated by the spike recovery, but spike recovery can be misleading when dealing with resistant forms. It has been shown that recovery of such forms is significantly improved, compared to Soxhlet or room-temperature extraction, by using methanol, acetone, or acetonitrile at 75 °C in a sealed vial, or by refluxing with methanol. Examples include field residues of EDB,[35] atrazine and metolachlor,[36] and simazine,[13] in addition to artificially generated resistant fractions of halogenated aliphatic hydrocarbons.[37] Both time and temperature are important.[36] Soxhlet is inferior apparently because the temperature inside the extraction thimble is only around 50–55 °C with these solvents.

3 Mechanism

Currently, the mechanism of retarded sorption is controversial. Depicted in Figure 1 are: (1) pore diffusion (PD), in which the rate-limiting step is molecular diffusion through fluids in particle micropores where advection does not occur, *i.e.* cracks and aggregate in interstices a few nanometers diameter; (2) the adsorption bond energy (ABE) mechanism, in which 'bond' making and breaking at the sorbent–water interface is rate-limiting (sometimes called chemical sorption); (3) intraorganic matter diffusion (IOMD), where molecules move slowly through a viscous polymeric SOM phase, analogous to the diffusion of small- to medium-size molecules

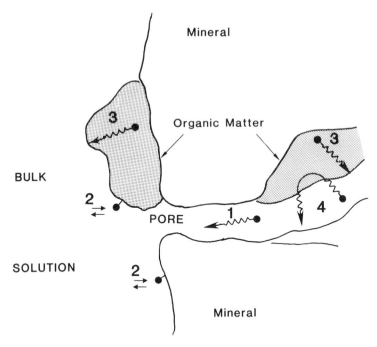

Figure 1 *Representation of several mechanisms for retarded sorption: (1) pore diffusion in fluids, PD; (2) adsorption bond energetics, ABE; (3) intra-organic matter diffusion, IOMD; and (4) sorption-retarded pore diffusion, SRPD*

through synthetic organic polymers; and (4) sorption-retarded pore diffusion (SRPD), in which micropore diffusion is retarded by rapid-reversible sorption to SOM on pore 'walls'.

Unretarded pore diffusion (PD) is usually not rate-limiting, as witnessed by the breakthrough curve symmetry widely observed for non-sorbing solutes (*e.g.* ^3H_2O, chloride) compared to sorbing solutes in soil columns. Moreover, there probably is insuffient microporosity in most materials to account for major effects.[18,38]

For non-ionic organic compounds, adsorption to mineral surfaces is believed to play a minor thermodynamic role in sorption except in materials of very low OC content or under very dry conditions. Although a contribution of adsorption to the kinetic component is conceivable, there is plenty of evidence that desorption of the resistant fractions is dependent on SOM. For example, the 'slowly reversible' fraction of EDB and TCE based on a particular sorption–desorption cycle was correlated with OC content, both among the different soils tested and for a given soil in which various amounts of SOM had been removed with hydrogen peroxide.[39] In another example,[24] the leachability according to particle size of predominantly resistant atrazine and metolachlor was practically constant when normalized to the OC content of each size fraction, suggesting that desorption had occurred from the SOM. Finally, inverse linear free-energy relationships (LFER) have been established between a desorption rate constant k_{-2} [eqn. (2)] and the K_p or K_{ow} (*i.e.* log–log plot was linear with a negative slope).[40–43]

Adsorption Bond Energy Mechanism

Many feel uncomfortable with the idea that non-polar or weakly polar organic compounds can form associations at the solution–sorbent interface that are strong enough to be rate-limiting. It is generally believed that the forward adsorption reaction requires a very small activation energy, while the reverse (desorption) is activated by at least the free energy of the interaction; *i.e.* the potential energy well in which the molecule resides at the surface. Non-ionic compounds can engage in van der Waals interactions, dipole–dipole interactions, π–π bonds between aromatic or highly conjugated groups, and hydrogen bonds. Individually, none of these are more than a few kcal mol^{-1}. Such bonds are formed or broken in very short times in solution. Thus, one might expect that microscale adsorption is fast on a practical time scale.

Recall, however, that desorption of hydrophobic compounds from DOM was in the order of hours[20] and that polar pesticide molecules were not completely desorbed from DOM in 4 d.[21] Dissolved humic acid molecules may approximate the smallest or most penetrable humic substances on soil particles.

Further interesting results have been obtained with surface-treated inert silica. Burris *et al.*[44] carried out studies of PCE and 1-methylnaphthalein (MN) on silica gel coated with humic acid. The coating was prepared by acidifying a suspension of 80–20 mesh silica in an alkaline solution of Aldrich humic acid and heating overnight at 100 °C. They observed asymmetric breakthrough curves (more so for MN), and very slow approach to equilibrium—lasting several days for PCE and up to 100 or more days for MN—in batch uptake experiments. Electron microscopy of the coated particles revealed that the

'... humic acid coating may be relatively thin and not present as isolated aggregates on the surface ... The micrographs did not indicate deep cracks leading to the interior of the grains, although distinct shallow crevasses were noted. Interpretation of the micrographs must be viewed with caution, as they can give only crude indication of the nature of the HA coating.'[44]

Szecody and Bales[45] found desorption time scales of up to hours during column transport of substituted benzenes on C_1, C_8, and C_{18}, and alkyl-phenyl modified silicas. They concluded that simple diffusion was not adequate to explain the slow rates, and suggested a chemical step. The slowest rates were on the alkyl-phenyl silica, suggesting to them π–π interactions between surface and sorbate aromatic rings.

Despite the above results, it is difficult to believe that adsorptive forces alone could account for the long-term behaviour that is now routinely observed. Such behaviour is displayed even by small, non-polar, moderately hydrophobic molecules like TCE or EDB. Additional arguments against ABE have been made by others.[40]

Sorption-retarded Pore Diffusion

SRPD assumes intraparticle radial diffusion kinetics in which an effective diffusion coefficient, D_{eff}, is of the form[10,18]

$$D_{eff} = D_w \kappa_r / \tau (1 + \beta K_p / \theta) \qquad (8)$$

where D_w is the bulk aqueous diffusivity (*ca.* 10^{-5} cm^2 s^{-1}); K_p is the partition coefficient for equilibrium sorption on pore walls; β and θ are the particle density and porosity, respectively; τ (≥ 1) is a tortuosity factor accounting for non-linear diffusion paths, dead-end pores, and variable pore diameters; and κ_r (≤ 1) is a term for pore constrictivity factor and steric effects at the molecular level. As the form of eqn. (8) indicates, diffusion in pore water thus may be slowed relative to bulk water by sorptive retardation [the $(1 + \beta K_p / \theta)$ term], tortuosity, and constriction.

The SRPD model is supported by results of halogenated hydrocarbon sorption in soils, sediments, and aquifer materials. In some instances the rate of approach to equilibrium increased with decreasing particle size, suggesting that particle diameter is the characteristic diffusion length scale.[10,18] Consistent with eqn. (2), rates were inversely correlated with K_p or K_{ow} of the compounds.[10,18] Finally, pulverization of the soil greatly accelerated uptake[17,18] or desorption.[28,39]

Intraorganic Matter Diffusion

Brusseau and Rao and co-workers[40–43,46] have presented several major arguments in favour of IOMD. They fit column elution curves to a two-compartment equilibrium first-order kinetic sorption model [eqn. (2)] coupled to the advection-dispersion equation [eqn. (1) without reaction]. Fits to this model afford the overall sorption constant (as K_p) and the desorption rate constant k_{-2}. Low OC soils and fairly rapid flow rates were used, meaning that contaminant residency times were in the order of minutes to a few hours. It is not clear, therefore, whether their results apply to long-term phenomena of the kind so far discussed. The value of k_{-2}, in fact, is quite sensitive to flow rate.[43]

Separate k_{-2} *vs.* K_p LFERs were obtained for substituted *s*-triazines on the one hand, and chloroaliphatics and substituted (—OH, —O$^-$, and —Cl) and unsubstituted aromatics on the other.[42] The triazine line was about two log units of k_{-2} lower than the line for the other groups, suggesting that polar interactions with SOM slowed the movement of the triazine molecules relative to the others. It was argued that the magnitude of this shift was more consistent with IOMD than SRPD. An inverse relationship between k_{-2} and OC content,[46] and a direct correlation between k_{-2} and fraction of methanol in the eluent[41] highlighted the importance of SOM for the kinetic component of sorption. The organic cosolvent was suggested to 'soften' the SOM polymer. Lastly,[40] as the chain length of *n*-alkylbenzenes increased, a point was reached where no further decrease in k_{-2} occurred, in accord with diffusion behaviour of *n*-alkanes in organopolymers, but not in accord with SRPD, the analogy for which is diffusion in zeolites.

Other results seem to be inconsistent with SRPD, lending support for IOMD in view of the significance of SOM. Desorption of field-weathered EDB,[28] and atrazine and metolachlor[24] were practically independent of particle size down to clay-size particles $(0.2–2\ \mu m)$, indicating that the characteristic diffusion length scale is much smaller than the particle radius. The upper-limit of D_{eff} was estimated to be in the order of 10^{-17} to $10^{-16}\ cm^2\,s^{-1}$. SRPD cannot be invoked, at least for these compounds, without requiring K_p to be many orders of magnitude greater than expected values of $\sim 10^1\ ml\,g^{-1}$, or calling upon severe tortuosity or steric hinderance [see eqn. (8)]. The elution curve of the non-sorbing penta-fluorobenzoate ion was identical to that of ^3H_2O, despite its considerably greater size, seemingly ruling out severe steric constriction.[40]

It is conceivable that IOMD and SRPD are operating simultaneously, or that one or the other dominates depending on soil structure. If further research shows that relatively long times (hours) are typical for desorption from DOM, then the rapid-reversible assumption of the SRPD model is incorrect. In that event, experimental distinction between IOMD and SRPD becomes difficult. Even the conceptual distinction between the two becomes fuzzy, depending on the extent to which SOM may be regarded as having an extended internal surface with a rigid micropore structure, as opposed to being a gel-like organic phase.[47,48]

4 Research Needs

There is ample evidence that sorption has a fast component, with equilibrium times in the order of hours, and a much slower component. Since K_p determinations are routinely carried out over 1 or 2 d, it is likely that many K_p values represent mainly the fast component rather than overall sorption. As discussed above, the fast component may be only a small percentage of the total. Furthermore, models predict[24] that changes in K_p (over days) after the initial rapid uptake can be mistaken as being negligible, particularly if changes in the liquid-phase concentration and not K_p itself are used to judge equilibrium. All of this raises questions about how far published K_p values are from their true values.

It is then fair to raise other thermodynamic issues. Models for many environmental processes use K_p as a parameter. Do we therefore need to re-evaluate those correlations? This would be unnecessary if the short-term K_p value was linear with K_{eq} and responded in the same way to external conditions. One study, however, showed that K for the equilibrium fraction was independent of ionic strength whilst K for the 'resistant' fraction was not.[19] Even with a strong K_p–K_{eq} correlation, processes occurring over intermediate time scales may be far more sensitive to rate than equilibrium constants. In the end, confidence in our models will be improved by working closer to equilibrium.

Have sorption kinetics been mistaken for other processes? Puzzling sorption behaviour appears from time to time; *e.g.* the solids' concentration effect. In addition, there is a continuing debate over adsorption *vs.* hydrophobic partitioning; the latter does not fully explain the often observed non-linearity of sorption isotherms and behaviour in low OC solids. To what extent does kinetics affect results and cloud interpretations?

Obviously, more kinetic data are needed. The goal is a workable, field relevant mathematical fate and transport model requiring the least experimental input. A mechanistic understanding of sorption is crucial to that goal. Kinetic data from DOM or model polymeric colloids will hopefully give insight as to what occurs at the particle–water interface. Diffusion constants at the single particle level—soil, surface-coated clay, of humic acid particles—may help resolve mechanistic issues. An electrodynamic thermogravimetric technique has been used to study organic vapour sorption on single particles of montmorillonite clay and commercial sorbents.[49] Studies using artificial media—for example, surface-modified silica or humic acid-coated silica—are likely to be useful. It is necessary to thoroughly evaluate the effects of component and environmental variables on sorption kinetics: compound structure, SOM content and composition, soil structure, temperature, pH, salinity, *etc.* Finally, ways must be found to overcome the obstacle that slow sorption represents to remediation efforts.

References

1. J.J. Pignatello, in 'Reactions and Movement of Organic Chemicals in Soils', eds. B.L. Sawhney and K. Brown, SSSA Special Publication No. 22, Soil Science Society of America, Madison, WI, 1989, Chap. 5, p. 45.
2. M.L. Brusseau and P.S.C. Rao, *CRC Crit. Rev. Environ. Cont.*, 1989, **19**, 33.
3. M. Alexander, *Soil Sci. Soc. Am.*, 1965, **29**, 1.
4. J.W. Hamaker and J.M. Thompson, in 'Organic Chemicals in the Environment', eds. C.A.I. Goring and J.W. Hamaker, Marcel Dekker, New York, 1972, Chap. 2, p. 49.
5. D.R. Cameron and A. Klute, *Water Resour. Res.*, 1977, **13**, 183.
6. S.W. Karickhoff, in 'Contaminants and Sediments', ed. R.A. Baker, Ann Arbor Science Publishers, Ann Abor, MI, 1980, Vol. 2, p. 193.
7. S.W. Karickhoff, *J. Hydraul. Eng.*, 1984, **110**, 707.

8. S.W. Karickhoff and K.R. Morris, *Environ. Toxicol. Chem.*, 1985, **4**, 469.
9. S.W. Karickhoff and K.R. Morris, *Environ. Sci. Technol.*, 1985, **19**, 51.
10. S. Wu and P.M. Gschwend, *Environ. Sci. Technol.*, 1986, **20**, 717.
11. D.I. Gustafson and L.R. Holden, *Environ. Sci. Technol.*, 1990, **24**, 1032.
12. J.J. Pignatello and L.Q. Huang, *J. Environ. Qual.*, 1991, **20**, 222.
13. S.L. Scribner, T.R. Benzing, S. Sun and S.A. Boyd, *Environ. Sci. Technol.*, 1992, **21**, 115.
14. S.G. Pavlostathis and G.N. Mathavan, *Environ. Sci. Technol.*, 1992, **26**, 532.
15. J.A. Smith, C.T. Chiou, J.A. Kammer and D.E. Kile, *Environ. Sci. Technol.*, 1990, **24**, 676.
16. J.J. Pignatello, *Environ. Toxicol. Chem.*, 1991, **10**, 1399.
17. W.P. Ball and P.V. Roberts, *Environ. Sci. Technol.*, 1991, **25**, 1223.
18. W.P. Ball and P.V. Roberts, *Environ. Sci. Technol.*, 1991, **25**, 1237.
19. J.M. Santana-Casiano and M. González-Dávila, *Environ. Sci. Technol.*, 1992, **26**, 90.
20. J.P. Hassett and E. Millicic, *Environ. Sci. Technol.*, 1985, **19**, 638.
21. D. Lee and W.J. Farmer, *J. Environ. Qual.* 1989, **18**, 468.
22. P.V. Roberts, M.N. Goltz and D.M. MacKay, *Water Resour. Res.*, 1986, **22**, 2047.
23. M.N. Goltz and P.V. Roberts, *J. Contam. Hydrol.*, 1988, **3**, 37.
24. J.J. Pignatello, F.J. Ferrandino and L.Q. Huang, *Environ. Sci. Technol.*, 1993, **27**, 1563.
25. A.V. Ogram, R.E. Jessup, L.T. Ou and P.S.C. Rao, *Appl. Environ. Microbiol.*, 1985, **49**, 582.
26. M.E. Miller and M. Alexander, *Environ. Sci. Technol.*, 1991, **25**, 240.
27. B.N. Aronstein, Y.M. Calvillo and M. Alexander, *Environ. Sci. Technol.*, 1991, **25**, 1728.
28. S.M. Steinberg, J.J. Pignatello and B.L. Sawhney, *Environ. Sci. Technol.*, 1987, **21**, 1201.
29. U. Varanasi, W.L. Reichert, J.E. Steion, D.W. Brown and H.R. Sanborn, *Environ. Sci. Technol.*, 1985, **19**, 836.
30. J.J. Pignatello, C.R. Frank, P.A. Marin and E.X. Droste, *J. Contam. Hydrol.*, 1990, **5**, 195.
31. S.R. Wild, K.S. Waterhouse, S.P. McGrath and K.C. Jones, *Environ. Sci. Technol.*, 1990, **24**, 1706.
32. A. Marcomini, P.D. Capel, T. Lichtensteiger, P.H. Brunner and W. Giger, *J. Environ. Qual.*, 1989, **18**, 523.
33. D.M. MacKay and J.A. Cherry, *Environ. Sci. Technol.*, 1989, **23**, 630.
34. US Environmental Protection Agency, Summary Report High-Priority Research on Bioremediation, Bioremediation Research Needs Workshop, Washington, DC, 1991, p. 7.
35. B.L. Sawhney, J.J. Pignatello and S.M. Steinberg, *J. Environ. Qual.*, 1988, **17**, 149.
36. L.Q. Huang and J.J. Pignatello, *J. Assoc. Off. Anal. Chem.*, 1990, **73**, 443.
37. J.J. Pignatello, *Environ. Toxicol. Chem.*, 1990, **9**, 1107.
38. W.P. Ball, C. Buehler, T.C. Harmon, D.M. MacKay and P.V. Roberts, *J. Contam. Hydrol.*, 1990, **5**, 253.
39. J.J. Pignatello, *Environ. Toxicol. Chem.*, 1990, **9**, 1117.
40. M.L. Brusseau, R.E. Jessup and P.S.C. Rao, *Environ. Sci. Technol.*, 1991, **25**, 134.

41. M.L. Brusseau, A.L. Wood and P.S.C. Rao, *Environ. Sci. Technol.*, 1991, **25**, 903.

42. M.L. Brusseau and P.S.C. Rao, *Environ. Sci. Technol.*, 1991, **25**, 1501.

43. M.L. Brusseau, *J. Contam. Hydrol.*, 1992, **9**, 353.

44. D.R. Burris, C.P. Antworth, T.B. Stauffer and W.G. MacIntyre, *Environ. Toxicol. Chem.*, 1991, **10**, 433.

45. J.E. Szecsody and R.C. Bales, *J. Contam. Hydrol.*, 1989, **4**, 181.

46. P. Nkedi-Kizza, M.L. Brusseau, P.S.C. Rao and A.G. Hornsby, *Environ. Sci. Technol.*, 1989, **23**, 814.

47. K.D. Pennell and P.S. Rao, *Environ. Sci. Technol.*, 1992, **26**, 402.

48. C.T. Chiou, J-F. Lee and S.A. Boyd, *Environ. Sci. Technol.*, 1992, **26**, 404.

49. L. Tognotti, M. Flytzani-Stephanopoulos and A.F. Sarofim, *Environ. Sci. Technol.*, 1991, **25**, 104.

7

Interactions of Organic and Inorganic Contaminants with Soils: Unifying Concepts

By Donald S. Gamble, Cooper H. Langford,[1] and G.R. Barrie Webster[2]

CLBRR, AGRICULTURE CANADA, RESEARCH BRANCH, OTTAWA, ONTARIO K1A 0C6, CANADA
[1]DEPARTMENT OF CHEMISTRY, UNIVERSITY OF CALGARY, 2500 UNIVERSITY DRIVE, CALGARY, ALBERTA T2N IN4, CANADA
[2]DEPARTMENT OF SOIL SCIENCE, UNIVERSITY OF MANITOBA, WINNIPEG, MANITOBA R3T 2N2, CANADA

1 Introduction

Thirty years or more ago it was universally assumed that it would never be possible to do rigorously quantitative chemistry with such complicated natural mixtures as agricultural soils, aquatic sediments, natural waters, or even their components. One of the practical implications of that assumption was that the physical chemistry interactions of metal ions and organic chemicals with the natural geochemical mixtures could never be understood or described quantitatively in terms of molecular level mechanisms. Practical chemical problems arising from the use of fertilizers and pesticides were consequently formulated in empirical or semi-empirical forms leading to practical limitations in soil and environmental chemistry. Empirical descriptions largely represent the conditions of the particular experiments out of which they have come, and support predictions of low reliability. The types of extrapolations and generalizations that make it possible to compare different samples or experiments and to do predictive calculations for practical hydrological engineering require exact mathematical descriptions of real processes or mechanisms. For such interactions as labile sorption and desorption, retarded intraparticle diffusion, and catalysed chemical reactions, mathematical descriptions based on chemical stoichiometry thus provide the best results.

   The research strategy (outlined below) has been derived from a sequence of investigations that followed a logical progression through a set of geochemical systems of increasing complexity approaching practical field conditions. In the first stage, many authors published physical and chemical properties of clays, humic materials, and hydrous metal oxides. The second stage consisted of investigations into the solution-phase reactions of fulvic acid with cations and organic pesticides. This led automatically to the work on cation and pesticide reactions with undissolved humic acid in two physical phases. In the fourth stage of the progression, work with humic materials in two physical phases is now providing both theoretical and experimental support for work with whole mineral soils. Research with aquatic sediments also belongs to this fourth stage of complexity. Progress can be determined by entering the results of laboratory studies of mechanisms into hydrological models, and then testing the computer predictions against field experiments.

Since the research strategy is not yet widely accepted and its benefits are only now being recognized, its nature and validity warrant a careful examination. Other authors have published excellent reviews of the mathematical methods for interpreting equilibria and kinetics in mixed geochemical systems[1-4] and the experimental methodology which forms the basis of this research strategy is discussed elsewhere.[5-7] The next step is to show that the whole research strategy is a success using both published and current research.

2 Outline of the Research Strategy

The creation and application of the concepts, mathematical descriptions, and to some extent the experimental methods, have been guided primarily by the research strategy. These components are therefore best seen within the context of the strategy. Although most of its components are separately familiar, their assembly into a coherently integrated research strategy is an important achievement; it is the means by which stoichiometrically exact chemistry can be done with complex geochemical mixtures such as soils, sediments, and natural waters in the same sense that exact chemistry is done with monomeric pure reagents. It is thus possible to produce mathematical descriptions of physical chemical mechanisms that can be used for predictive calculations.

The research strategy is as follows:

(i)   Mathematical description of a mechanism is formulated in terms of three types of information:
    a.   constants and variables that can be measured or theoretically deduced;
    b.   constants and variables that provide the required useful answers;
    c.   mathematical equations that state the behaviour and constraints of the system under investigation.
    Calculation provides the means to go from convenient measurements to the required parameters.

(ii) Mixtures of non-identical reactive sites or binding sites are treated as chemical systems, rather than as large numbers of separate components.

(iii) The concept of weighted averages is applied to the law of mass action for equilibria in the natural mixtures of geochemical systems. A similar form of the concept has been applied to chemical kinetics in the mixed systems.

(iv) The theory of categories of variables is used for the design of experiments, interpretive and predictive calculations, and the reporting of research results.

(v) Categories of reactive or binding sites are identified when possible.

(vi) Chemical units are used for mass and concentration in the quantitative descriptions of material balances, equilibrium, and chemical kinetics; these units are a prerequisite to the application of chemical stoichiometry.

(vii) Total numbers of reactive or binding sites are measured as capacities or saturation limits. The resulting numerical values are used in defining the stoichiometries of equilibrium and kinetics calculations.

(viii) Natural mixed systems are chemically characterized in terms of particular chemical reactions or binding processes, by means of experimental scans across the components of the mixtures, *e.g.* by titration.

(ix) Experimental methods have been developed with which free, labile bound, and relatively non-labile bound chemical species may be identified, and tracked kinetically. Mechanisms can thereby be quantitatively described.

3 Categories of Variables

Three categories of variables are distinguished from each other, for the conduct, calculation, and molecular level interpretation of environmental chemistry systems.[8] They are described below.

(i) Inner variables: Inner variables are those identified by theory as being required to specify the state of a chemical system. They should be used for correlations, molecular level interpretative calculations, and predictive calculations. Typical examples include the numbers or concentrations of protonated carboxyl groups, and the mole fraction of sorption sites occupied by a sorbed pesticide.

(ii) Outer variables: Outer variables are those that are manipulated operationally for the conduct of an experiment. Typical examples include pH, total moles of pesticide or metal ion added to a sample, and the ratio of weight of solid sample to volume of experimental solution. They are more practical than inner variables for setting up and running many types of experiments; frequently inner variables cannot be used for that purpose.

(iii) Background descriptive variables: These variables are useful for developing qualitative insight or for making practical decisions. They are not used in chemical calculations, and do not appear in the mathematical descriptions of the system. Typical examples include elemental analysis, and per cent organic matter in a mineral soil.

Incorrect use of the three types of variables can cause unnecessary data scatter and some confusion. For example, hydrogen-bond structure formation[6,9] might cause a sorption rate constant to be a function of the mole fraction of the sorption sites covered by pesticide molecules. A plot of the rate constants for several soils against pH would likely show excessive data scatter. At one pH, different soils would have different numbers of protonated carboxyl groups and, therefore, different mole fractions of their sites occupied. The outer variable, pH, would be an indirect reflection of the correct inner variable, the mole fraction of the sorption sites occupied. If pH were used for empirical black box correlations, the resulting fitted curve would represent only the particular experiment from which it had come. Because it would not represent the actual molecular level phenomena, it would lack the generality necessary for good predictive calculations. It is therefore important to compare experimentally measured dependent variables to the proper independent variable and to be aware of the danger of using a variable from the wrong category.

Descriptions of equilibrium and kinetic theory and a detailed discussion of the analytical chemistry based on stoichiometry and speciation are described at length in a review paper currently being prepared for publication.

4 The Success of the Research Strategy

The research strategy has produced a continuous stream of experimental successes which are presented in the order of both increasing complexity and increasing practical significance.

(i)   Burch *et al.*[8] published some of the earliest proofs that exact predictive calculations can be done with a system as complicated as a fulvic acid solution. The solid curve in Figure 1 gives the predicted behaviour of the weighted average equilibrium function for its proton dissociation. The calculations were based on previously published theory. The solid dots and squares were direct experimental measurements. The authors likewise demonstrated that dilution effects on the equilibrium can also be predicted.

(ii) Buffle[1] assembled data for Cu^{2+} complexed by humic materials. He found that even a partial adherence to the principle that correlated variables should directly represent cause and effect relationships enabled him to plot equilibrium data from many sources on the same curves. A much more complicated humic–metal ion system

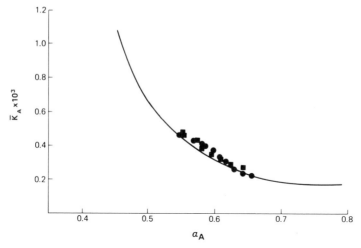

Figure 1 *The acid dissociation of fulvic acid carboxyl groups: an experimental test of theory. The solid line shows the predicted behaviour of the weighted average equilibrium function, as the degree of ionization increases. The circles and squares are experimental measurements for two different concentrations of fulvic acid*

yielded an especially significant demonstration that classical chemical behaviour can be observed in such cases. Gamble *et al.*[9] developed a rigorous theoretical description of mixed equilibrium experiments in which 11 metal ions competed for the mixture of cation-exchange sites in an undissolved humic acid. Empirically tabulated experimental data failed to exhibit the classical Irving–Williams series of complexing stability constants for the metal ions. It was obscured by a combination of competing equilibria and the averaging inherent in the macroscopic measurements made on a whole mixture. A simple type of calculation based on the theory produced differential equilibrium functions that revealed the anticipated Irving–Williams series.

(iii) The acid catalysed hydrolysis of atrazine in fulvic acid solutions provided a striking example of the stoichiometrically exact chemistry that can be done with complicated mixtures. Gamble and Khan[6] found that the correct choice of variables makes it possible to simplify the graphical presentation of the rate constant data. Figure 2 is the conventional but impractical plot against pH. Each of the non-linear curves was produced by a different concentration of bulk fulvic acid. Functional group concentrations have been ignored in this conventional plot. Consequently no curve in this graph can be used to predict any other curve. The numerical values for the rate constant expression on the *y* axis have in fact been caused by the protonated carboxyl groups carried by the fulvic acid, and not

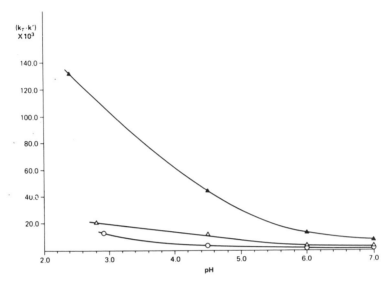

Figure 2 *The rate constant plot for atrazine hydrolysis catalysed by the carboxyl groups of dissolved fulvic acid at 25 °C. The use of the outer variable pH produces a family of non-linear curves that do not go through the origin*

by the humic polymer itself. Logic therefore says that the rate constant data should be plotted against the molarity of the causative protonated carboxyl groups. Figure 3 proves this to be correct. All of the same experimental data are now on a single straight line that goes through the origin which can be used for predictive calculations.

(iv) Atrazine hydrolysis in systems with two physical phases has given some evidence that suggests generality. The open squares in Figure 4 represents kinetic experiments for atrazine hydrolysis that has been catalysed by the carboxyl groups in undissolved humic acid.[10] A similar experiment conducted with peat soil gave the data plotted with solid circles.[11] Within the limits of experimental error, all of the rate constants seem to have the same dependence on sorption site loading. Not only is this further evidence for the effectiveness of the research strategy, but also it should make another type of predictive calculation possible. Figure 4 suggests the possibility of predicting the hydrolysis rate constant from measurements of sorption site coverage.

(v) Predicted intraparticle diffusion of atrazine in a mineral soil has been checked against experiment.[12] Equation (1) has been derived from a theoretical model developed by Crank.[13]

$$\ln(\theta_L) = \ln(A) + Z\ln(t) \qquad (1)$$

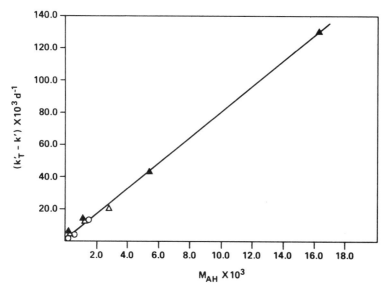

Figure 3 *The atrazine hydrolysis kinetics experiments of Figure 2, replotted against the inner variable, molarity of protonated carboxyl groups. These carboxyl groups are the main catalyst but are not properly accounted for in Figure 2*

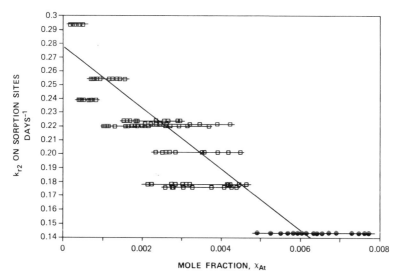

Figure 4 *Atrazine hydrolysis kinetics in systems with two physical phases. Open squares, humic acid suspensions. Solid circles, peat soil. χ is the inner variable, mole fraction of occupied sorption sites. Its use gives a plot that indicates that the two quite different samples follow the same trend*

In this model, steady-state labile surface sorption drives the retarded intraparticle diffusion into a semi-infinite sink. θ_L is the mol g^{-1} of atrazine taken up by intraparticle diffusion; A is a constant related to the diffusion coefficient and the geometry of the surface normal to the direction of travel of the difffusing molecules; and t is time in seconds. The test of the theory is that an experimental plot of $\ln(\theta_L)$ *vs.* $\ln(t)$ should give a slope of $Z = 1/2$, if the process is diffusion. A set of experiments like that in Figure 5 has given the slopes plotted in Figure 6. The two histogram bars on the right-hand end show a very close agreement with theory and the experimental mean. In addition, almost all of the individual experiments agreed with the theoretical value of $1/2$, to within the standard errors (a more demanding criterion than standard deviations would be). If Crank's model applies to atrazine in a mineral soil, then the intraparticle diffusion rate should have a linear dependence on the extent of surface sorption. Excellent experimental support for the model is found in Figure 7. In this figure, θ_{At} is the labile surface sorption. The research strategy has now matured sufficiently so that molecular level mechanisms can be determined.

(vi) The ultimate test of the research strategy is to check the laboratory research results against field experiments. While much work remains, a few preliminary cases are available. The solid curve in Figure 8 shows the extent to which some laboratory research currently in progress agrees with a field experiment published by

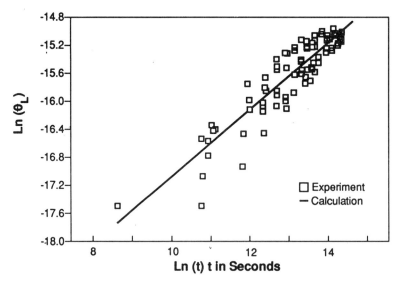

Figure 5 *The test for intraparticle diffusion, adapted from the steady-state surface sorption model of Crank.[13] See Equation (1). An atrazine–mineral soil was used.* Z = 0.4899, Standard Error = 0.0229. Ln(A) = −22.0195, Standard Error = 0.234

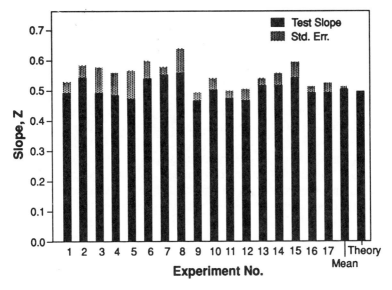

Figure 6 *A set of intraparticle diffusion tests, like that in Figure 5. Atrazine–mineral soil experiments were used*

Bowman.[14] In both instances, the concentrations of atrazine remaining in solution are shown. Sorption and intraparticle diffusion caused the losses. Even though this graph does not account for the effects of water as a solvent, the laboratory experiment and the field measurements are not seriously divergent. The neglect of the water has been overcome by the hydrological modelling of Clemente,[15] in Figure 9. The laboratory research conducted according to the concepts and methods of the research strategy has identified the mechanism for the atrazine–soil interaction. It has also calibrated the equilibrium and kinetics parameters of the mechanism. Clemente used this information as input for the hydrological model. Similar checks against field measurements were obtained for each of three summers supporting the validity of the research strategy. Pignatello[16] has independently made suggestions about mechanisms that are similar to the mechanism that has been experimentally identified and used here with a hydrological model.

5 Practical Implications

Environmental chemistry research is valuable in contributing to the solution of existing problems and in preventing future problems. A new multidisciplinary technology based on this research will enable predictive engineering calculations. The documented successes of the research strategy imply that it will permit this to be achieved for the chemical portion of the multidisciplinary technology. An important implication is

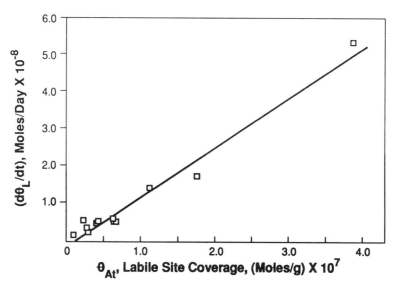

Figure 7 *The dependence of the intraparticle diffusion rate on labile surface site coverage at 25 °C. Atrazine in a mineral soil at 25 °C. The diffusion rate drops to zero when there is no surface sorption on the soil particles*

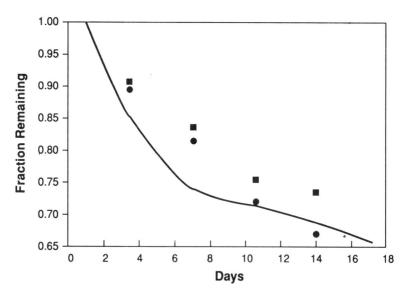

Figure 8 *A comparison of laboratory and field experiments, for atrazine in mineral soils. Solid line, from the high-performance liquid chromatography (HPLC)–microfiltration method in the laboratory. Solid circles, Bowman's field experiment showing only atrazine.[14] Solid squares, Bowman's field experiment, showing atrazine + desethylatrazine[14]*

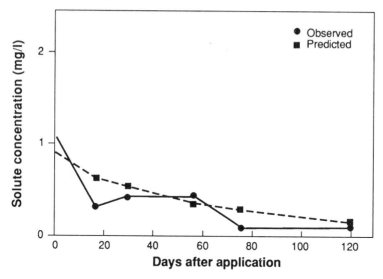

Figure 9 *A comparison of predicted and observed atrazine concentrations in the field.*[12,13] *Solid squares; hydrology model predictions by Clemente using a quantitative description of the chemical mechanism developed in the laboratory. Solid circles; field measurements by Clemente*[15]

that once the concepts and methods have been set in place, the production of a much larger data base will become possible.

References

1. J. Buffle, 'Complexation Reactions in Aquatic Systems: An Analytical Approach', Ellis Horwood, Chichester, 1988, Chap. 6, pp. 314–322.
2. J. Buffle and R.S. Altman, 'Aquatic Surface Chemistry', ed. W. Stumm, Wiley Interscience, New York, 1987, Chap. 13, pp. 351–383.
3. C.H. Langford and D.W. Gutzman, *Anal. Chim. Acta*, 1992, **256**, 183.
4. M.S. Shuman, B.J. Collins, P.J. Fitzgerald and D.L. Olson, 'Aquatic and Terrestrial Humic Materials', eds. R.F. Christman and E.T. Gjessing, Ann Arbor Science Publishers, Ann Arbor, MI, 1983, Chap. 17.
5. D.S. Gamble and S.U. Khan, *Can. J. Chem.*, 1992, **70**, 1597.
6. D.S. Gamble and S.U. Khan, *Can. J. Soil Sci.*, 1985, **65**, 435.
7. Z. Wang, D.S. Gamble and C.H. Langford, *Anal. Chim. Acta*, 1991, **244**, 135.
8. R.D. Burch, C.H. Langford and D.S. Gamble, *Can. J. Chem.*, 1978, **56**, 1196.
9. D.S. Gamble, M. Schnitzer, H. Kerndorff, and C.H. Langford, *Geochim. Cosmochim. Acta*, 1983, **47**, 1311.
10. D.S. Gamble and S.U. Khan, *Can. J. Chem.*, 1988, **66**, 2605.
11. D.S. Gamble and S.U. Khan, *J. Agric. Food Chem.*, 1990, **38**, 297.
12. D.S. Gamble, J. Li, G. Gilchrist and C.H. Langford, in preparation.
13. J. Crank, 'The Mathematics of Diffusion', Oxford University Press–Clarendon Press, Oxford, 1975.
14. B.T. Bowman, *Environ. Toxicol. Chem.*, 1990, **9**, 453.

15. R.S. Clemente, 'A Mathematical Model for Simulating Pesticide Fate and Dynamics in the Environment (PESTFADE)', Ph.D. Thesis, MacDonald Campus, McGill University, Montréal, Quebec, Canada, 1992.

16. J. Pignatello, in 'Reactions and Movement of Organic Chemicals in Soils', eds. B.L. Sawhney and K. Brown, SSSA Special Publication No. 22, Soil Science Society of America, American Society of Agronomy, Madison, WI, 1989, Chap. 3.

8

Sub-surface Transport of Natural Organic Matter: Implications for Contaminant Mobility

By John F. McCarthy

ENVIRONMENTAL SCIENCES DIVISION, OAK RIDGE NATIONAL LABORATORY, OAK RIDGE, TENNESSEE, USA

1 Introduction

The transport of contaminants in the sub-surface is controlled by both chemical factors related to the sorption of the contaminant to immobile surfaces of the aquifer and by hydrologic factors that determine the rate of flow of water within the aquifer. Typically, contaminant transport is described as a two-phase system, with a mobile phase composed of groundwater and a solid phase composed of the immobile aquifer surfaces. Chemicals with limited solubility in water, or with a high affinity for binding to either charged surfaces of minerals or to organic coatings on the aquifer media, are generally regarded as being relatively immobile in the sub-surface. For many sub-surface environments, where the concentrations of natural organic matter (NOM) in the groundwater are very low (>0.2 mg C l^{-1}) and there are no stable inorganic colloids present in the mobile groundwater phase, this paradigm is reasonable. However, the presence of NOM or inorganic colloids in soil or groundwater can compete with sorption sites on aquifer surfaces and effectively increase the apparent solubility of contaminants in the mobile groundwater phase, thus enhancing the mobility of the contaminant[1] (Figure 1). In this paper, NOM is defined as the total dissolved organic matter or colloidal organic matter in groundwater, and includes both humic and non-humic material.

The significance of NOM to the sub-surface transport of contaminants depends on several factors. First, the effect of NOM on the overall partitioning of a contaminant among the solid, aqueous, and NOM phases must be evaluated; that is, does the contaminant have a high affinity for binding to NOM?

Secondly, is the NOM mobile over significant distances in the aquifer? Alternately, will the NOM sorb to the aquifer and reduce the transport of the contaminant by increasing the organic carbon content (and thus the

Two-Phase System

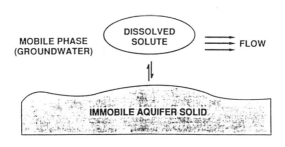

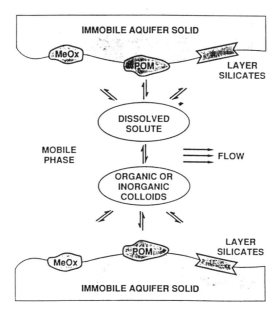

Figure 1 *The potential role of mobile colloids in enhancing the transport of highly adsorbing contaminants is illustrated. When colloids are present, consideration must be given to the distribution of contaminants between the aquifer, colloid, and groundwater, in addition to the interactions of the colloids with each other (agglomeration) or with the aquifer surfaces (filtration). POM and MeOx refer to particulate organic matter and metal oxide surfaces, respectively*
[Reprinted from ref. 2 with permission from *Environmental Science and Technology*]

adsorptive capacity) of the porous media? Adsorption of NOM to inorganic colloidal (submicron-sized) components of the aquifer can disperse colloids such as iron oxide or clays within the aquifer. The NOM can promote the electrostatic or colloidal stability of inorganic colloids and enhance the transport of the colloids, and of any adsorbed contaminants.

Finally, the heterogeneity of NOM must be recognized. NOM is a complex mixture of solutes, macromolecules, and probably aggregates of those components that differ in size, charge and hydrophobicity (see Chapters 2 and 3). Components of NOM differ with respect to both binding of organic and inorganic contaminants and sorption onto aquifer surfaces. In this sense the key questions concerning the role of NOM in contaminant transport can be described as; 'Which sub-components of NOM bind contaminants?' and 'Are those sub-components mobile in porous media?'

This paper addresses those questions, and considers the evidence from laboratory and field studies related to the potential significance of NOM to the sub-surface transport of contaminants. The transport of NOM, and of sub-components of NOM, is considered first, then the implications of NOM transport to the mobility of contaminants is discussed. Finally, the role of NOM in the stability and transport of inorganic colloids, and of associated contaminants, is evaluated. Throughout this paper, the behaviour of NOM is discussed in terms of the predictable properties of operationally defined sub-components of NOM, which include both humic and non-humic components of non-particulate organic matter in ground-water systems.

2 Adsorption and Transport of NOM

Research has shown that NOM may be immobilized through complex interactions with mineral surfaces.[2-6] Information on the potential immobilization of NOM by soils or sub-surface sediments is more limited. Sibanda and Young[5] demonstrated the adsorption of NOM on iron-rich tropical soils, while Leenheer[7] demonstrated that hydrophobic components of NOM had a higher affinity for soils relative to the hydrophilic components. Jardine *et al.*[8] reported that NOM was highly reactive with forest soils even under conditions of extreme preferential flow. The mechanism of NOM retention by soils and sediments is largely unknown. Tipping[4] proposed ligand exchange of surface coordinated OH^- and H_2O from iron oxides by humic substances. Jardine *et al.*[6] suggested that a portion of the NOM was bound to soils *via* anion exchange, but that the predominant mechanism was by physical adsorption mechanisms driven by favourable entropy changes, as evidenced by low heats of adsorption, and the preferential adsorption of hydrophobic components of NOM.

Most mechanistic studies of adsorption or immobilization of NOM by soils or aquifer materials have utilized batch adsorption techniques. Although batch techniques allow individual adsorption mechanisms to be evaluated, it is not clear that batch results can be directly extrapolated to the dynamic flow conditions found in groundwater due to the presence of multiple adsorption mechanisms, localized variability in solution chemistry, and non-equilibrium sorption conditions along flow paths. Observations of NOM mobility in flowing systems is perhaps of more direct relevance to evaluating the role of NOM in contaminant transport in sub-surface systems, and will be the focus of the discussion below. Although the transport of NOM through the vadose zone has been examined, especially with respect to the flux of nutrients through watersheds,[9-11] this paper focuses on NOM transport within saturated systems.

NOM Transport in Laboratory Columns

The transport of NOM from a wetlands pond was investigated in well-packed laboratory columns containing sub-surface sediments from a shallow sandy coastal aquifer.[12] The aquifer media was approximately 90–95% sand, with 3–10% clay, 0.3–4.1 mg g^{-1} total iron, and 0.03–0.05% organic carbon. Breakthrough curves (BTC) for the total NOM at a range of concentrations are shown in Figure 2. Key observations include:

(i) While the bulk of the NOM is retarded to breakthrough of a non-reactive tracer (Br$^-$), approximately 10–20% of the NOM co-elutes with the unretarded tracer. However, additional experiments confirmed that there was no significant size-exclusion processes operative in transport of the NOM through the sediment; that is, there was no evidence that macromolecular components of NOM were excluded from smaller pore structures, as had been observed for transport of the organic macromolecule, blue dextran (2 000 000 daltons relative molecular mass[13]).

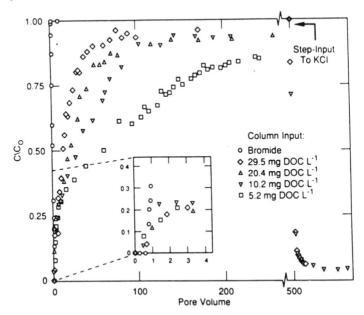

Figure 2 *Observed breakthrough curves for Br$^-$ and various concentrations of NOM transported through laboratory columns containing aquifer material from Georgetown, South Carolina, with C/C$_0$ representing a reduced concentration*
[Reprinted from ref. 12 with permission from *Soil Science Society of America*]

(ii) The NOM BTCs reflect adsorption by the porous media. The extensive tailing is attributed to non-linearities in the NOM adsorption isotherms,[6,14] time-dependent adsorption during transport, and the existence of multiple adsorption processes during transport [see (v), below].

(iii) The maximum for non-equilibrium adsorption was achieved more rapidly at higher input concentrations of NOM, reflecting the mass of NOM needed to saturate the adsorptive capacity of the column media.

(iv) Concentrations of NOM in the column effluent declined rapidly when NOM inputs were terminated (step input to KCl solution without NOM), suggesting that desorption of NOM adsorbed to the media is very slow.

(v) There were compositional changes in the NOM as it passed through the column, reflecting multiple adsorption processes and the heterogeneity in the adsorptive properties of sub-components of NOM.

To quantify the preferential adsorption of NOM sub-components along the flow path, the absorptivity of the column effluent was measured. Absorptivity of the NOM at 260 and 330 nm (a_{260} and a_{330}, respectively) increased significantly and slowly approached steady state above 200 pore volumes. The ratio of a_{260}/a_{330} also changed significantly during the NOM displacement experiments, suggesting significant compositional changes of the NOM during transport. The hypothesis of preferential adsorption and 'chromatographic separation' of different NOM sub-components was evaluated *via* column transport experiments using isolated sub-components of NOM prepared by the XAD-8 fractionation technique of Leenheer.[15] Hydrophilic (Hl)–NOM eluted more rapidly than did either the hydrophobic–acid (HbA) or hydrophobic–neutral (HbN) fractions of NOM (Figure 3). Heterogeneity of adsorptive properties was apparent in the transport of even the isolated XAD-8 fractions, as evidenced by increases in a_{260} of both the HbA and HbN fractions during the displacement experiments. Higher adsorptivity at 260 nm has been associated with increased aromatic content of the NOM,[16,17] suggesting that more aromatic, and probably more hydrophobic, NOM components have a higher affinity for adsorbing to the column matrix.[18]

The results of the column transport studies demonstrated that NOM can be mobile in soil systems, that the extent of retardation is significantly different for different sub-components of NOM, but that continued input of NOM can saturate the binding sites on the porous media and permit the unretarded transport of even the more reactive NOM sub-components.

Field Scale Mobility of NOM in a Sandy Aquifer

Observations on the transport of NOM in laboratory columns provide no reliable insight into how these processes (measured in the laboratory) are expressed at the field scale. To have confidence in predictions of NOM

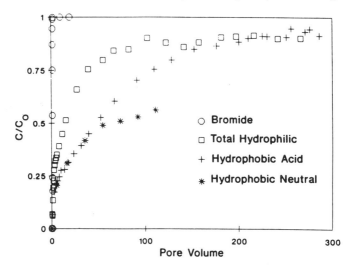

Figure 3 *Observed breakthrough curves for the isolated XAD-8 fractions of NOM transported through laboratory columns containing aquifer material from Georgetown, South Carolina, with C/C_0 representing a reduced concentration*
[Reprinted from ref. 12 with permission from *Soil Science Society of America*]

transport, field experiments are essential. Discrepancies between predictions of NOM mobility *vs.* observed breakthrough reveal new processes that are only expressed at the field scale. These potentially important processes or factors include:

(i)   misrepresentation of the chemical kinetics of adsorption, possibly due to slower flow and longer contact times with porous media in field studies;

(ii) spatial variability in important chemical properties of the aquifer material. Such variability may create differential sorption zones and enhance preferential transport paths; or

(iii) variability in porosity, creating physical zones of preferred flow.

To address this concern, the mobility of NOM and its sub-components was measured in a natural aquifer to determine if the key features of NOM retention by aquifer material at the field scale (1.5–3 m) was consistent with laboratory batch studies and in the small, 8 cm long soil columns. These comparisons are crucial to describing and predicting the role of NOM in contaminant transport in aquifers.

A field injection experiment was performed in the sandy coastal aquifer in Georgetown, South Carolina that was the source of the aquifer material used in the laboratory column experiments. The NOM was from a wetlands pond. The NOM contained 37% Hl–NOM, 42% HbA–NOM, and 21% HbN–NOM. A large volume (80 000 l) of the pond water with

66 mg l^{-1} of NOM was injected into an anoxic, iron-rich sandy aquifer. The forced gradient was continued after the NOM injection using groundwater with low levels of NOM. Sampling wells were located in three horizons at distances 1.5 and 3 m from the injection well (Figure 4). A non-reactive tracer (Cl^{-}) was injected with the frist 4000 l of DOC solution to provide information on the average pore water velocity and dispersion characteristics of the media.[19,20]

Despite heterogeneity problems in describing the field scale transport of the non-reactive tracer,[21] the results of the NOM injection revealed similarities in addition to differences in NOM transport at the spatial scales of the laboratory *vs.* field. Basic features of NOM breakthrough observed in laboratory column studies (extended tailing to long times and rapid decline in concentrations of NOM in the mobile phase when NOM inputs were terminated) were observed in the field experiment (Figure 5). Retardation of NOM in the field was in general agreement with that predicted from laboratory batch studies and column studies.

Differences at the two scales were also evident. The unretarded breakthrough of 10–20% of the NOM observed in the laboratory columns was only observed in the A3 well within the most hydraulically transmissive zone with high pore water flow velocities (Figure 4). Transit times for the other zones were slow enough to permit sorptive interactions to dominate.

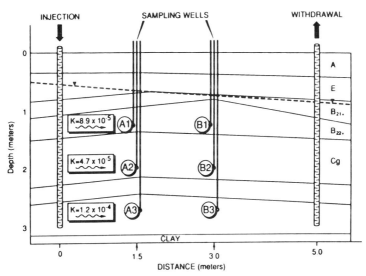

Figure 4 *Cross-sectional diagram of the forced gradient dipole well array at Hobcaw Field, Georgetown, South Carolina. The hydraulic conductivity (K, in* m s^{-1}*), estimated from a two-dimensional flow model,[21] for the 1.2, 2.25, and 2.7 m horizons is indicated. The soil designations for the profile are indicated on the right of the figure*
[Reprinted from ref. 19 with permission from Lewis Publishers, Chelsea, MI]

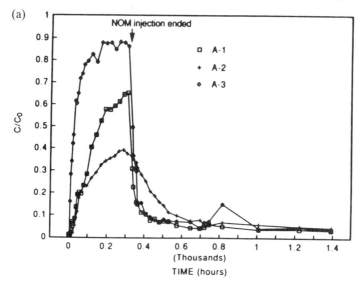

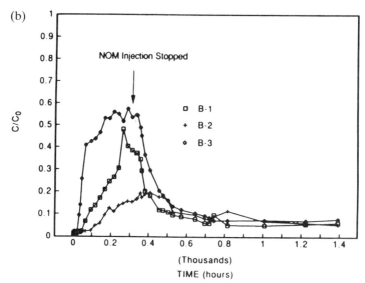

Figure 5 *Concentration histories of total organic carbon concentrations (expressed as reduced concentrations, C/C_0) in groundwater recovered from (a) A-wells and (b) B-wells for 1400 h following initiation of the NOM injection. The injection of the NOM solution was terminated at 312 h*
[Reprinted from ref. 19 with permission from Lewis Publishers, Chelsea, MI]

An important contribution of this study was to confirm that key chemical processes responsible for adsorption of different sub-components of NOM are operative at the field scale. In spite of the physical and chemical heterogeneities in the aquifer,[19-21] and the much longer flow paths and time scales involved in the field experiment, clear differences were seen in the mobility of size fractions and XAD-8 fractions of NOM. As in the laboratory columns, hydrophobic NOM was retarded in the aquifer, but continued inputs of NOM eventually saturated sorption sites on the aquifer to permit essentially unretarded transport of even the more-adsorptive fractions. While that process could require more time at lower concentrations of NOM, the results are significant in demonstrating that the maximum sorption capacity of the aquifer was achieved in spite of the complexity of hydrological, geochemical, and microbiological processes operative at the field scale.

3 Implications of NOM Transport to Contaminant Mobility

In general, more is known about the interaction of NOM with contaminants than is known about the mobility of NOM in either the vadose or saturated-zone environments. Humic substances from surface water or soil bind hydrophobic organic contaminants. This interaction can enhance the contaminant's apparent solubility in water, and reduce its apparent affinity for binding to sediment particles and to the immobile phase of reversed-phase chromatographic columns and soil columns.[18,22] The polyelectrolytic character of humic and fulvic acids also enables them to associate with metal ions. Laboratory and field studies have demonstrated the significance of humics on the speciation of metals[23] and radionuclides.[24] Humic and fulvic acids from soil and sediment bind actinides strongly.[25] Colloidal organic matter from lakes also binds radionuclides and inhibits their adsorption to sediments.[24]

The importance of NOM to contaminant transport can be varied and depends on several conditions that are likely to be site-specific. Several aspects of these issues will be considered below.

Effect of Mobile NOM on Contaminant Transport

Classical convective-dispersive (CD) transport models are based on equilibrium or kinetic reactions of a contaminant between an immobile (solid) phase and a single mobile (solution) phase, and do not consider the possible influences of a second mobile colloid or NOM phase (see Figure 1). One approach used to account for multiple mobile sorbents has been to reparameterize the CD equation to include size-exclusion processes. This reparameterized CD equation has been used to describe the transport of macromolecule-associated contaminants in soil columns.[13,26] However, results of Abdul *et al.*[27] and Dunnivant *et al.*[12] using naturally occurring NOM demonstrated that size exclusion was not occurring in the transport of NOM through soil columns. Kan and Tompson[28] considered the influence of a second mobile phase on the transport process by refining the distribution coefficient of the classical CD equation. This approach allows a contaminant to be distributed between an immobile solid phase, a mobile solution phase, and a mobile sorbent phase (such as NOM). Such a description is an essential step toward our understanding of and predicting the effect of NOM on contaminant transport in the sub-surface. Unfortunately, Kan and Tompson[28] used only model organic sorbents (Triton X-100 and bovine serum albumin) to test their approach to facilitated transport of organic contaminants. More recently, Dunnivant *et al.*[18] used a similar reformulation of the CD equation to test predictions of facilitated transport of an organic and inorganic contaminant using NOM and aquifer material from the field site in Georgetown, South Carolina, described earlier.

The one-dimensional transport of a single solute in porous media can be described by the CD equation:

$$\rho/\theta(\partial S/\partial t) + \partial C/\partial t = D(\partial^2 C/\partial X^2) - V(\partial C/\partial X) \tag{1}$$

where ρ is the porous medium bulk density, θ is the volumetric water content, S is the total adsorbed solute per unit mass of solid, t is time, C is the resident concentration of solute in the mobile phase, D is a dispersion coefficient reflecting the combined effects of diffusion and hydrodynamic dispersion on transport, X is distance, and V is the mean pore water velocity. Typically, the contaminant distribution between the solid and solution phase is represented by an equilibrium distribution coefficient, K_d. When NOM is present as an additional component of the mobile phase, the equations can be modified to account for the additional phase:

$$K_d = S/C_{aq} \tag{2}$$

$$K_{app} = S/(C_{aq} + C_{NOM}[NOM]) \tag{3}$$

$$K_{app} = K_d/(1 + K_{NOM}[NOM]) \tag{4}$$

$$R = 1 + (\rho/\theta)K_{app} \tag{5}$$

$$R = 1 + (\rho/\theta)(K_d/1 + K_{NOM}[NOM]) \tag{6}$$

where K_{app} (apparent K_d) represents the equilibrium distribution of the contaminant in the presence of mobile NOM, S is the solid-phase contaminant concentration (mg g^{-1} solid), C_{aq} is the aqueous-phase contaminant concentration (mg ml^{-1} solution), C_{NOM} is the concentration of contaminant associated with NOM (mg g^{-1} carbon), [NOM] is the concentration of NOM (mg carbon ml^{-1}), K_{NOM} (= C_{NOM}/C_{aq}) is the equilibrium distribution of the contaminant between NOM and water in the absence of the solid phase, and R is the net retardation factor, which is an expression of the mean velocity of the contaminant compared to that of the water flowing through the porous media.

Combining eqns. (1) and (5) yields the general form of the transport equation:

$$R(\partial C/\partial t) = D(\partial^2 C/\partial X^2) - V(\partial C/\partial X) \qquad (7)$$

Selection of either eqn. (2) or (4) [for use in eqn. (7)] will depend on the presence or absence of mobile NOM.

This formulation was evaluated using columns containing aquifer material for organic contaminants [two polychlorinated biphenyls (PCB)], and the inorganic contaminant, cadmium. In this system, the K values were used to represent the distribution of both contaminants, recognizing the mechanistic differences between partitioning of PCBs and complexation of cadmium. For both PCBs (Figure 6) and cadmium, contaminant mobility was found to increase as the solution NOM concentrations were incrementally increased. Experimental adsorption BTCs were predicted independently of column experiments using contaminant distribution coefficients (K) measured using batch techniques and the CD transport equation [eqn. (7)]. The observed and predicted BTCs agreed well, especially for the PCB (Figure 6). Although the absolute increases in cadmium mobility could not be predicted using the batch data set, relative increases in R values did allow the co-transport of cadmium by NOM relative to the reference BTC (no NOM) to be predicted. Results supported the hypothesis that contaminants can be co-transported by mobile NOM in groundwater systems.[18]

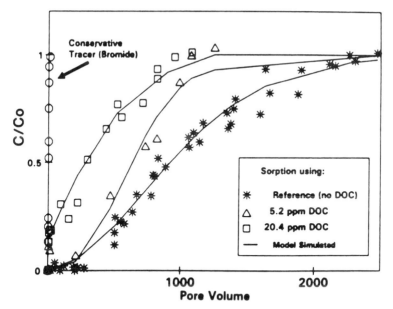

Figure 6 *Breakthrough of 2,2',4,4',5,5'-hexachlorobiphenyl from laboratory columns containing aquifer material from Georgetown, South Carolina, in the presence of different NOM concentrations. C/C_0 represents a reduced concentration. All elution waters contained 0.025 M KCl and 0.0002% NaN$_3$ (by weight). Solid curves represent model predictions [Eqn. (7)] obtained using batch data*
[Reprinted from ref. 18 with permission from *Environmental Science and Technology*]

Changes in NOM Composition Along Flow Paths: Implications for Contaminant Transport

In the experiments of both Kan and Tompson[28] and Dunnivant *et al.*,[18] the porous media was in contact with the NOM solutions until adsorptive equilibrium appeared to be achieved. Thus, these studies considered only the distribution of the contaminant with solution, NOM, and solid phases, and were able to ignore simultaneous interactions of NOM with the solid phase along the flow path through the column. This is a reasonable scenario for environmental conditions with continuing inputs of NOM over long times. However, in many situations, NOM will not be in equilibrium with respect to adsorption onto the porous media along a flow path. Failure to consider the interaction of the 'mobile' NOM with the porous media can result in large errors in predicting the mobility of contaminants, due to:

(i)   decreasing concentrations of NOM along a flow path as the NOM adsorbs to the solid phase (*i.e.* a change in [NOM]);

(ii) preferential adsorption of sub-components of NOM that have differing affinities for specific classes of contaminants (*i.e.* a change in K_{NOM}); or

(iii) alteration of the surfaces of the porous media due to adsorption of 'mobile' NOM (*i.e.* a change in K_d).

Concentrations of NOM in effluents of soil columns or in sampling wells down gradient of input sources of NOM are initially lower than the concentration of NOM applied to the column or injected into the aquifer. The concentration of mobile NOM gradually increases as binding sites are saturated.[12,19,20,27] The resulting increase in the organic carbon content (OC) of the porous media can increase retardation of contaminants that bind to organic matter. For example, continued inputs of high concentrations of NOM in laboratory columns (100–400 pore volumes of NOM at $\approx 60\ \mathrm{mg\,l^{-1}}$) increased the OC of the aquifer sediments from 0.039 to 0.058%.[18] This change had no measurable effect on the K_d for binding of the PCBs to the sediment, since sediment organic content appears to significantly influence binding of organic contaminants only when OC exceeds 0.1%.[29] The mobility of cadmium was, however, highly dependent on the OC of the solid phase; higher OC resulting from longer contact with the NOM solutions increased the time necessary to achieve steady-state breakthrough of the cadmium in the soil columns.[18]

Less well understood is the effect of compositional changes in NOM resulting from preferential adsorption of some NOM sub-components. In both laboratory and field experiments, smaller, more hydrophilic NOM was more mobile than were larger, more hydrophilic components[12,19,20] (Figure 3). This result has clear implications for contaminant transport in groundwater because different components of NOM differ in their capacity to bind, and potentially co-transport contaminants. For example, in a study of 11 surface- and ground-waters, there was a good correlation between the size (molar volume) and HbA content of the NOM in the water sources and the K_{NOM} of those sources for binding of the polycyclic aromatic hydrocarbon (PAH), benzo(*a*)pyrene[30] (BaP). The capacity of either natural waters[30] or of isolated humic material[16,17] to bind PAHs is well correlated with the UV absorbance (a_{270}) and aromatic content of the NOM.

Isolated XAD-8 fractions of NOM differ significantly with respect to their capacity to bind PAHs and PCBs. However, the binding capacity of the unfractionated NOM can be described as the weighted average of the binding capacity of individual fractions[31,32] (Figure 7). Isolated Hl–NOM had little affinity for either class of organic contaminant. Both HbN– and HbA–NOM had a high affinity for binding BaP; however, the capacity to bind several PCBs resided almost exclusively in the HbN fraction[31,32] (Figure 6).

Although it appears to have a minor role in binding of hydrophobic organic contaminants, Hl–NOM may be very significant to the transport of particle-reactive metals and radionuclides. Binding of particle-reactive cations to NOM is generally understood to be related to the total exchangeable acidity of the NOM.[33] Hl–acid–NOM is higher in oxygen-to-carbon ratios and enriched in carboxylic functional groups,[34] relative to fulvic acids which are known to have a high affinity to bind metals and radionuclides.[25,33,35] It is, therefore, quite likely that this non-humic fraction of NOM would have a significant capacity to complex metal and radionuclide contaminants.

Both field and laboratory results confirm that NOM is mobile in groundwater, and sub-components of NOM differ in their mobility within the sub-surface, with small, hydrophilic components of NOM displaying more mobility than larger hydrophobic components. This suggests that NOM may be more effective in enhancing the mobility of metals and radionuclides which form complexes with acidic functionalities on the Hl–NOM; however, more research is needed to examine the role of the non-humic, hydrophilic components of NOM in binding, and possibly co-transporting contaminant cations.

Although initially retarded, larger hydrophobic NOM capable of binding and co-transporting hydrophobic organic contaminants travels with little retardation after continued contact with the aquifer matrix. Although the Georgetown field experiment used NOM concentrations that were very high, and the average groundwater flow velocities induced by the forced gradient were fast compared to normal flow even through sandy aquifers, the results are relevant to NOM facilitated transport under conditions that are more representative of those of typical groundwater. For example, even much lower concentrations of NOM (5 mg carbon l^{-1}) increased the mobility of a PCB and cadmium in soil columns using materials from the field site.[18] The field experiment, by confirming the laboratory descriptions of NOM transport, suggests that NOM would have been effective in enhancing the field-scale transport of contaminants.

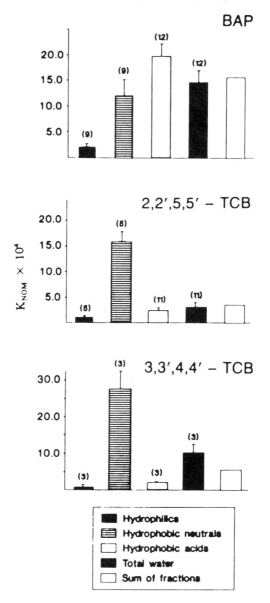

Figure 7 *Distribution coefficients* (K_{NOM}) *of the total (unfractionated) water sample from a wetland pond in Hyde County, North Carolina and different NOM fractions for BaP, and two PCBs, 2,2',5,5'-tetrachlorobiphenyl (TCB) and 3,3',4,4'-tetrachlorobiphenyl. Bars indicate the mean values (± SD) and the number in parentheses is the number of replicates. The measured K_{NOM} for the total water agrees very well with a cumulative K_{NOM} ('sum of fractions') calculated from the sum of the K_{NOM} for the individual fractions and the relative contribution of each XAD fraction to the total NOM*
[Reprinted from ref. 32 with permission from Lewis Publishers, Chelsea, MI]

Effects of NOM on the Mobility of Inorganic Colloids and Associated Contaminants

There is an increasing body of evidence that components of the inorganic solid phase may exist in groundwater in the colloidal (submicron) size range.[1,36] These colloidal particles have similar composition and surface characteristics to the immobile aquifer solids, but may be mobile within aquifers. These mobile sorbents represent a potentially important vector for transport of metals through the subsurface, although the conditions controlling their mobility are poorly understood. Recent work has attempted to include the role of a colloidal phase in speciation models[37] (COMET). However, it is more difficult to predict the extent to which colloids will be transported in sub-surface environments because little, if any, information is available on the abundance and distribution of colloidal particles in groundwater, or the hydrogeochemical conditions controlling their formation and mobility in sub-surface systems.

NOM may be a critical factor in maintaining the negative surface potentials of newly formed or dispersed particles and limiting the deposition of colloids to aquifer materials (generally negatively charged). For example, colloids such as iron oxides carry a small net positive charge near neutral pH;[38] thus they are inherently unstable from a colloid stability point of view and deposition on aquifers is favoured. In both laboratory studies[39] and studies of natural surface waters,[2,4,40,41] the association of NOM with iron oxides has been demonstrated to result in an overall negative surface potential at pH 6.5 and to an increase in the colloid stability. Ryan and Gschwend[42] postulated that colloidal hydrous oxides of Fe, Al, and Ti in a coastal sedimentary aquifer were stabilized as suspensions in groundwater by coatings of organic carbon on the inorganic particles.

Liang *et al.*[43,44] observed the formation of stable iron oxide colloids resulting from oxygenation of Fe(II) in an anoxic, iron-rich aquifer in Georgetown, South Carolina, during the NOM injection experiment described earlier in this paper. In the zone within the aquifer where dissolved oxygen substantially increased from 0.1 to 1.5 mg l^{-1}, turbidity increased by a factor of 10. The increased turbidity was associated with a decrease in Fe(II) and increase in Fe(III) due to oxidation, with most of the Fe(III) in the colloidal size range (> 100 nm). Although most of the NOM was < 3000 daltons molecular weight (approximately 1 nm), approximately 5% of the organic carbon was retained on the 100 nm filter, suggesting that it was associated with the colloidal iron (hydr)oxide. The electrophoretic mobilities of groundwater particles in the oxygenated zone showed negative surface potentials (-20 mv).[43,44] At the pH of the groundwater (pH 6.0–7.2), most iron (hydr)oxide particles have a positive or neutral charge.[38] The consistently negative surface potential from electrophoretic mobility measurement suggests that iron colloids are associated with anions. Fourier transform infrared spectroscopy analysis on the groundwater showed a strong spectrum for —COO⁻ (N. Marley[45]); hence, the association of iron colloids with organic anions may result in the negative surface potential of particles.

Laboratory experiments have demonstrated that oxidation products of Fe(II) formed in the presence of NOM concentrations > 0.5 mg carbon l^{-1} are stable and are efficiently transported through quartz porous media. Iron colloids formed in the presence of lower NOM concentrations are less stable and very few of the colloids are transported through laboratory columns (L. Liang *et al.*[46]).

Thus, even relatively low concentrations of NOM may promote the transport of contaminants that are associated with inorganic colloids (including oxides and hydroxides of actinide elements such as uranium, neptunium, plutonium, and americium) by promoting the colloidal stability of the colloids and permitting their transport through sub-surface systems.

4 Summary and Conclusions

Although it adsorbs to mineral surfaces, significant amounts of NOM can potentially be transported through soil and aquifer material. At both the laboratory and field scale, preferential adsorption of some sub-components of NOM has been demonstrated, and suggests greater retardation of more hydrophobic components of NOM, relative to hydrophilic components. Even at the field scale, however, continued inputs of the NOM can saturate binding sites in the porous media and permit transport of the more adsorbable components. These observations are significant to contaminant transport because different sub-components of NOM bind different types of contaminants to varying extents. Furthermore, NOM may facilitate the transport of contaminants that bind preferentially to mineral surfaces by promoting the colloidal stability of inorganic colloids that would normally have a positive or neutral surface charge at the groundwater pH.

Given the chemical and physical heterogeneity of NOM, and variability in those properties among different sources of NOM, it seems unlikely that any generalized capability to predict NOM behaviour (with respect to either NOM transport, contaminant sorption, or colloidal stability) will arise from site-specific descriptions of the behaviour of the total organic material. Rather, the behaviour of the total NOM must be viewed as the weighted average of the predictable properties of sub-components of NOM. Research is needed to determine the most meaningful ways of fractionating NOM to address specific activities or behaviour. Sub-components can be fractionated according to the key properties hypothesized to be relevant to certain behaviour, be it hydrophobicity, abundance of ionizable functional groups, or the acidic strength of the NOM. Furthermore, research is needed to understand the behaviour of non-humic components of NOM because these materials appear to be important to NOM transport and to binding of some contaminants.

Acknowledgements

The author is supported by the Sub-surface Science Program, Environmental Sciences Division, Office of Health and Environmental Research, US Department of Energy (US DOE). The Oak Ridge National Laboratory (ORNL) is managed by Martin Marietta Energy Systems, under Contract No. DE-AC05-84OR21400 with the US DOE Publication No. 4047 of the Environmental Sciences Division of ORNL.

References

1. J.F. McCarthy and J.M. Zachara, *Environ. Sci. Technol.*, 1989, **23**, 496.
2. J.A. Davis, *Geochim. Cosmochim. Acta*, 1982, **46**, 2381.
3. J.A. Davis and R. Gloor, *Environ. Sci. Technol.*, 1981, **15**, 1223.
4. E. Tipping, *Geochim. Cosmochim. Acta*, 1981, **45**, 191.
5. H.M. Sibanda and S.D. Young, *J. Soil Sci.*, 1986, **37**, 197.
6. P.M. Jardine, N.L. Weber and J.F. McCarthy, *Soil Sci. Soc. Am. J.*, 1989a, **53**, 1378.
7. J.A. Leenheer, 'Contaminants and Sediments, Volume 2, Analysis, Chemistry, Biology', ed. R.A. Baker, Ann Arbor Science Publishers, Ann Arbor, MI, 1980, p. 267.
8. P.M. Jardine, G.V. Wilson, R.J. Luxmoore and J.F. McCarthy, *Soil Sci. Soc. Am. J.*, 1989b, **53**, 317.
9. W.H. McDowell and T. Wood, *Soil Sci.*, 1984, **137**, 23.
10. P.M. Jardine, G.V. Wilson, J.F. McCarthy, R.J. Luxmoore, D.L. Taylor and L.W. Zelazny, *J. Contam. Hydrol.*, 1990, **6**, 3.
11. G.F. Vance and M.B. David, *Geochim. Cosmochim. Acta*, 1992 in the press.
12. F.M. Dunnivant, P.M. Jardine, D.L.Taylor and J.F. McCarthy, *Soil Sci. Soc. Am. J.*, 1992, **56**, 437.
13. C.G. Enfield, G. Bengtsson and R. Lindqvist, *Environ. Sci. Technol.*, 1989, **23**, 1278.
14. P.M. Jardine, F.M. Dunnivant, H.M. Selim and J.F. McCarthy, *Soil Sci. Soc. Am. J.*, 1992, **56**, 393.
15. J.A. Leenheer, *Environ. Sci. Technol.*, 1981, **15**, 578.
16. T.D. Gauthier, W.R. Seitz and C.L. Grant, *Environ. Sci. Technol.*, 1987, **21**, 243.
17. S.J. Traina, J. Novak and N.E. Smeck, *J. Environ. Qual.*, 1989, **19**, 151.
18. F.M. Dunnivant, P.M. Jardine, D.L. Taylor and J.F. McCarthy, *Environ. Sci. Technol.*, 1992, **26**, 360.
19. J.F. McCarthy, L. Liang, P.M. Jardine and T.M. Williams, in 'Manipulation of Groundwater Colloids for Environmental Restoration', eds. J.F. McCarthy and F.J. Wobber, Lewis Publishers, Chelsea, MI, 1993, pp. 35–39.
20. J.F. McCarthy, T.M. Williams, L. Liang, P.M. Jardine, A.V. Palumbo, L.W. Jolley, L.W. Cooper and D.L. Taylor, *Environ. Sci. Technol.*, 1993, **27**, 667–676.
21. J. Mas-Pla, T.-C. Jim Yeh, J.F. McCarthy and T.M. Williams, *Ground Water*, 1992, **30**, 958–964.
22. D.E. Kile and C.T. Chiou, in 'Water Solubility Enhancement of Non-ionic

Organic Contaminants', eds. I.H. Suffet and P. MacCarthy, American Chemical Society, Washington, DC, 1989, p. 131.

23. J. Buffle, in 'Metal Ions in Biological Systems', ed. H. Siegel, Marcel Dekker, New York, 1984, p. 165.

24. D.M. Nelson, W.R. Penrose, J.O. Karttunen and P. Mehlhaff, *Environ. Sci. Technol.*, 1985, **19**, 127.

25. K. Nash, S. Fried, A.M. Friedman and J.C. Sullivan, *Environ. Sci. Technol.*, 1981, **15**, 834.

26. C.G. Enfield and G. Bengtsson, *Ground Water*, 1988, **26**, 64.

27. A.S. Abdul, T.L. Gibson and D.N. Rai, *Environ. Sci. Technol.*, 1990, **24**, 328.

28. A.T. Kan and M.B. Tompson, *Environ. Toxicol. Chem.*, 1990, **9**, 253.

29. R.P. Schwartzenbach and J.C. Westall, *Environ. Sci. Technol.*, 1981, **15**, 1360.

30. J.F. McCarthy, L.E. Roberson and L.E. Burris, *Chemosphere*, 1989, **19**, 1911.

31. J. Kukkonen, J.F. McCarthy and A. Oikari, *Arch. Environ. Contam. Toxicol.*, 1990, **19**, 551.

32. J. Kukkonen, J.F. McCarthy and A. Oikari, in 'Organic Substances and Sediments in Water', ed. R.A. Baker, Lewis Publishers, Chelsea, MI, 1991, p. 111.

33. E.M. Perdue, in 'Aquatic Humic Substances: Influence on Fate and Treatment of Pollutants', eds. I.H. Suffet and P. MacCarthy, American Chemical Society, Washington, DC, 1989, p. 281.

34. G.R. Aiken, D.M. McKnight, K.A. Thorn and E.M. Thurman, *Org. Geochem.*, 1992, in the press.

35. J.A. Marinsky and J. Ephraim, *Environ. Sci. Technol.*, 1986, **20**, 349.

36. J.F. McCarthy and C. Degueldre, in 'Environmental Particles', eds. H.P. van Leeuwen and J. Buffle, Lewis Publishers, Chelsea, MI, 1993, Chapter 6, p. 247.

37. W. Mills, B.S. Liu and F.K. Fong, *Ground Water*, 1991, **29**, 199.

38. R.O. James and G.A. Parks, 'Surface and Colloid Science', ed. Matijivic, Plenum, New York, 1982, Vol. 12.

39. L. Liang and J.J. Morgan, *Aquatic Sci.*, 1990, **52**, 32.

40. K.A. Hunter and P.S. Liss, *Nature (London)*, 1979, **282**, 823.

41. E. Tipping and D.C. Higgins, *Colloids Surf.*, 1982, **5**, 85.

42. J.N. Ryan, Jr. and P.M. Gschwend, *Water Resour. Res.*, 1990, **26**, 307.

43. L. Liang, J.F. McCarthy, T.M. Williams and L. Jolley, 'Manipulation of Groundwater Colloids for Environmental Restoration', eds. J.F. McCarthy and F.J. Wobber, Lewis Publishers, Chelsea, MI, 1993, p. 263.

44. L. Liang, J.F. McCarthy, L.W. Jolley, J.A. McNabb and T.W. Mehlhorn, *Geochim. Cosmochim. Acta*, 1993, **57**, 1987.

45. N. Marley, Argonne National Laboratory, personal communication.

46. L. Liang J.A. McNabb, J.M. Paulk, B. Gu and J.F. McCarthy, *Environ. Sci. Technol.*, 1993, in the press.

9

Modelling Solute–Sorbent Interactions of Saturated Flow in Heterogeneous Soils

By Göran Bengtsson

DEPARTMENT OF ECOLOGY, CHEMICAL ECOLOGY AND ECOTOXICOLOGY, UNIVERSITY OF LUND, HELGONAVÄGEN 5, S-223, 62, LUND, SWEDEN

1 Introduction

The ability to predict solute transport in a soil is limited by the interaction between mathematical model development and characterization of the features of the porous medium and the underlying physical, chemical, and biological processes. For non-reactive solutes in a homogeneous soil, models based on diffusion and advection give a mechanistic understanding of the transport. The spatial heterogeneity exhibited in the field can be described by a stochastic approach by which flow and flux distribution are random space functions, or by mass transfer of solute between mobile and immobile domains. None of these approaches requires a description of the structure of the porous medium, but different simplifications are necessary. A single sorption coefficient, unique for a combination of sorbent and sorbate and predicted from, *e.g.* sorbate aqueous solubility and sorbent organic carbon content, may be used to estimate retardation of reactive sorbates when local sorption equilibrium can be assumed. Asymmetric breakthrough curves are indicative of sorption non-equilibrium, resulting from chemical non-equilibrium or intrasorbent diffusion. Several types of mathematical models have addressed these problems and those of microbial degradation by simulating sorption–desorption kinetics and enzyme-reaction kinetics. Their merits depend on the understanding of the processes being modelled and the ability to design experiments and provide data for the model parameters.

2 Physical Heterogeneity and Sorption Non-equilibrium

Two of the major challenges to the current modelling of solute transport in soil, *viz.* physical heterogeneity and sorption non-equilibrium, can be addressed at the microscopic and the macroscopic scale. Whereas the microscopic scale allows discretization of transport variables within the representative elementary volume, the macroscopic scale is associated with the use of statistical averages of geometry and transport processes. The estimation of statistical variables becomes a problem itself at the macroscopic scale when the solid particle and the interstitial pore at the microscopic scale is replaced by low-permeability zones and continuous channels with high hydraulic conductivity, respectively. Yet, there are similarities between some of the conceptual approaches to model solute transport at the two scales. Most commonly, the non-homogeneous, non-equilibrium phenomena of solute transport are described by a two-phase approach. Physical heterogeneity is conceptualized as a mobile/immobile phase phenomenon, and sorption non-equilibrium is referenced as a fast/slow diffusion problem or an instantaneous/kinetic reaction process.

Physical heterogeneity at the microscopic level is largely attributed to bimodality of the distribution of the interstitial pore fluid velocity. The pore space is conceptually divided into a mobile phase, where solute transport is dominated by advection and dispersion, and a stagnant phase, *e.g.* narrow pores and boundary layers of particles, in which transport is limited by diffusion processes. The exchange of solute between the mobile and stagnant phases is most conveniently described as a first-order rate process:

$$J = \alpha(C_m - C_s) \qquad (1)$$

where J is the diffusion-like flux of solute between the mobile and stagnant fluid phases, α is a first-order mass transfer coefficient, which can be interpreted as a diffusion coefficient divided by an average diffusional distance, and C_m and C_s are average solute concentrations in the mobile and stagnant phase, respectively. The rate coefficient can be expressed as a diffusion coefficient[1] to give a more mechanistic explanation of the physical process and allow independent measurements of it. The applicability of the bimodality of pore fluid velocity for solute transport at the microscopic scale has been demonstrated[2-4] employing curve-fitting techniques.

Geochemical properties of the solid phase are considered a major determinant of sorption non-equilibrium phenomena. An implicit assumption in most sorption studies on organic solutes is that mineral clays and organic polymers are uniformly composed and distributed across the soil particles, so that sorbent characteristics within an elementary volume can be lumped into a physical average. Although the relationships between geochemical properties and sorption are poorly understood, there is some evidence for the importance of variations in soil particle mineralogy and particle size on sorption of organic solutes.[5-8]

The most obvious effect is due to particle size variations. The inverse relationship between particle size and sorption is due to two mechanisms (Figure 1). One is entirely a hydrodynamic effect as a result of the correspondence between particle size and hydraulic conductivity. The other one is a geochemical effect. In general, small particles, such as fine sand

and silt, which have a relatively large surface area, are dominated by magnetic minerals with a positive zero point of charge (ZPC), whereas larger particles are composed of minerals, *e.g.* quartz and feldspar, with negative ZPC (Table 1 in Barber *et al.*[8]). Dissolved organic material (DOM) will preferentially sorb to particles or sections of a particle with a positive ZPC as a result of complexation with phenolic and carboxylic groups[9,10] and subsequently facilitate solute sorption.

3 Sorption Models

Patchy distribution of organic carbon and clay minerals on soil particles has been documented in early work in soil science, and some recent data are available to support the correlation between the fraction of organic carbon on particles and sorption of hydrophobic and neutral solutes.[11-14] Sorption may also be influenced by the composition of the geopolymers, *e.g.* their aromaticity and O/C ratio.[14,15-18] These geochemical inhomogeneities are probably the underlying property of the solid material resulting in at least two distinct reaction patterns in batch sorption experiments, one fast and instantaneous followed by one slow and extended.[19,20] This type of behaviour has been conceptualized as compartment models, the most popular being a two-compartment or two-site model. It is assumed that the potential sorption or exchange sites are divided into those for which sorption is instantaneous and face the solute in parallel with those that are kinetically controlled[21] (Figure 2):

$$S_1 = fkC \qquad (2)$$

$$S_2 = (1 - f)kC \qquad (3)$$

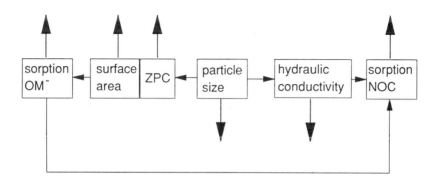

Figure 1 *Solute sorption increases indirectly due to reduced hydraulic conductivity and more directly due to an increased zero point of charge and surface area as the particle size becomes smaller*

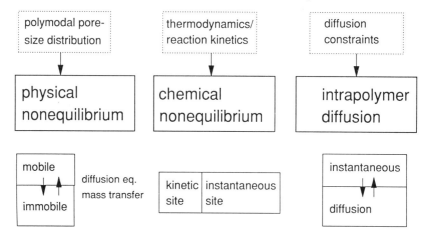

Figure 2 *Rate Determining Mechanism. Solute transport kinetics is conceptually understood as derived from a polymodal pore-size distribution, chemical reaction kinetics, and diffusion constraints in soil aggregates. The mobile/immobile phase representation of the physical heterogeneity includes a transfer function expressed by diffusion equations or a mass-transfer coefficient*

where S is the sorbed concentration, subscripts 1 and 2 refer to equilibrium and kinetic sites, respectively, C is the solution concentration, f is the fraction of sites assumed to be at equilibrium, and k is an empirical distribution coefficient. The mass balance of inputs and outputs for the sorbed concentration is given by the time derivatives, and assuming that sorption rates are controlled by first-order kinetics, where α is the first-order rate coefficient:

$$\partial S_1/\partial t = fk(\partial C/\partial t) \tag{4}$$

$$\partial S_2/\partial t = \alpha[(1 - f)\,kC - S_2] \tag{5}$$

The two-site, first-order kinetics model has been used with the linear convection-dispersion equation to describe asymmetrical breakthrough curves (BTCs) for various solutes in laboratory displacement experiments under steady-state flow conditions.[22-26] The concept has also been extended and modified to address a serial two-step sorption process including intrapolymer diffusion.[27] The existence of kinetic reactions of different rates controlling sorption at both sites has been also suggested,[19] and models describing sorption at three parallel sites[28] or a continuous distribution of sites[29] have been presented. Selim[30] proposed a model assuming that solute retention was due to a combination of (i) reversible and irreversible sorption, (ii) kinetic and equilibrium type of reactions, and (iii) fast and slow kinetic reactions (Figure 3). The author used the model to demonstrate that the consecutive kinetic reaction was necessary to describe breakthrough curves for P in various soils.

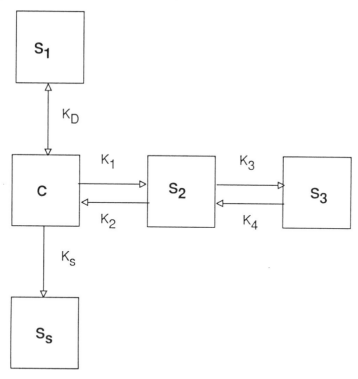

Figure 3 *A multi-compartment model for solute sorption reactions in a soil. The aqueous phase concentration (C) is in equilibrium with a solid-phase concentration (S_1) and reversibly associated by rate kinetics with loosely surface bound (S_2) and more firmly soil matrix bound (S_3) concentrations. Some of the solute will be irreversibly bound (S_s). K_1 to K_4 and K_s are first-order rate coefficients and K_D is a linear distribution coefficient (from Selim[30])*

A different approach was taken by Akratanakul *et al.*,[31] who considered the sorption process as a three-step mechanism. Solutes carried in a vertical flow direction were assumed to diffuse horizontally through the bulk solution to the sub-surface or boundary layer of the soil particle, diffuse laterally across the boundary layer, and finally react by, *e.g.* ion exchange, in an adsorbed layer at the particle surface. The mathematical description of the system required an experimental determination of seven variables, and their system was simplified by assuming that the boundary layer was thin and diffusion across it rapid in comparison with sorption reaction kinetics. The governing equations derived were for the bulk solution

$$\theta \frac{\partial C}{\partial t} = \frac{\partial}{\partial t}\left(\theta D_z \frac{\partial C}{\partial z}\right) - q_w \frac{\partial C}{\partial z} - \rho_b \frac{\partial S}{\partial t} \qquad (6)$$

and for the adsorbed layer

$$\rho_b \frac{\partial S}{\partial t} = \left(\frac{1}{\dfrac{1}{K_d} + \dfrac{1}{k_1 \zeta^{-b}}} \right) \theta C - \left(\frac{1}{k_1 \zeta^{-2b} \dfrac{1}{k_2} \dfrac{1}{K_d} + \dfrac{1}{k_2 \zeta^{-b}}} \right) \rho_b S \quad (7)$$

where θ is the volumetric water content; D_z is the dispersion coefficient in the vertical (z) direction; q_w is the Darcian flux in the z direction; ρ_b is the soil bulk density; $K_d = k_x/l$ where k_x is the horizontal mass transfer coefficient; l is the thickness of the bulk solution; k_1 and k_2 are the forward and backward reaction rate coefficients, respectively; ζ is the dimensionless equivalent to S; and b is a dimensionless constant associated with the rate of change of energy of adsorption as a function of surface coverage. Hence, solute diffusion across the boundary layer is characterized by the lumped coefficient K_d, and the chemical reaction kinetics by k_1, k_2, and b.

Akratanakul *et al.*[32] used the model with appropriate simplifications to describe sorption of Cd in unsaturated soil columns at varying pore water velocities. Values for K_d, k_1, and k_2 were calculated by curve-fitting analyses and found to be functions of the pore water velocity. Both the diffusion rate and the reaction kinetics k_1/k_2 increased with increasing water velocity, the former as a result of a decreased thickness of the boundary layer, and the latter due to a continuous supply of introduced Cd ions and a continuous removal of desorbed ions of various species. The experiments indicated that diffusion and chemical reactions proceed simultaneously and that the diffusion coefficient, regardless of pore water velocity, was sufficiently small to suggest that diffusion across the boundary layer was a rate-limiting step in sorption.

There are two kinds of problems associated with the correspondence between the sorption non-equilibrium models and the actual physical, chemical, and biological processes at the microscale level. One problem is to derive independent measurements of the semiempirical variables f, α, K_d, k_1, and k_2. They tend to vary with the pore water velocity and must often be fitted to observed data. Since sorbing solutes can be constrained by diffusive mass transfer resulting in asymmetrical BTCs not unlike the ones resulting from sorption non-equilibrium, it is difficult to distinguish between the contribution of each of the processes to non-equilibrium behaviour by curve fitting to experimental data alone. This is also evident from the equivalence of the solutions to the mathematical models for the two processes as long as sorption is linear and at equilibrium at one of the two kinds of sorption sites.[4,19] Brusseau *et al.*[33] developed a model to account for both physical and chemical non-equilibrium during solute transport in porous media and showed that the prediction of experimental BTCs in aggregated, organic soils improved over that of the two-site model when model variables were estimated from independent sources.

4 Nature of Interactions

Experimental approaches to resolve the nature of interactions include manipulation of the organic carbon content of sorbent and solvent. This was done by Nkedi-Kizza *et al.*[34] to demonstrate the influence of intrapolymer diffusion as a rate-limiting step in sorption. Asymmetric breakthrough curves (BTCs) resulting from miscible displacement of two herbicides in saturated soil columns could be attributed to sorption non-equilibrium. By plotting BTCs for the sorbing solutes, normalized for retardation, along with that for tritiated water after displacement in a soil with 0.2% organic carbon content (OC) and <0.01% OC after treatment with H_2O_2, the authors were able to demonstrate that sorption non-equilibrium was associated with the organic carbon content. The first-order rate coefficient for sorption obtained by fitting the two-site sorption–solute transport equation to the BTCs for the herbicides was inversely related to the organic carbon content of the soil. This suggests that intrapolymer diffusion was the rate-limiting sorption step, since the reduction of sorbent organic carbon content should reduce the path length for diffusion and hence increase the value of the rate coefficient.

However, the exact mechanisms of sorption non-equilibrium are still unresolved, and the relative contribution of diffusion and chemical reactions to the asymmetric BTCs at this microscopic level is an open question. A more detailed theoretical approach, analogous to that relating molecular topology of solutes to potential sorption,[35] to the distribution of exchange sites, sorbent 'hot spots', and diffusion layers, may improve the description of solute transport at the microscopic level and facilitate experimental differentiation of involved processes. Proper independent measurements of kinetic rate coefficients will remain a challenge to the experimenter. Agitated batch systems have been used with some success to derive solute–sorbent equilibrium distribution coefficients to predict retardation factors in dynamic column experiments, but the high solution-to-soil ratio and high contact rates are not representative for solutes moving through a soil, especially not an unsaturated one.

An alternative technique for studying sorption kinetics is a stirred continuous flow system[36,37] in which the solute and sorbent are mixed by stirring in a small solution volume under constant flow conditions. The major advantage of this system is that conditions are created to distinguish between instantaneous (local equilibrium) reactions and fast reaction kinetics with reaction half times of the order of one minute or less. However, the reaction kinetics under most experimental conditions are appropriately described by apparent rate coefficients including diffusion and chemical reaction kinetics, since diffusion as a rate-controlling mechanism is difficult to eliminate without introducing abrasion phenomena when true soils are used as sorbents. Most studies of sorption kinetics on soil material suggest that diffusion cannot be excluded as a rate-limiting mechanism, neither experimentally nor conceptually.[38–40]

5 Deterministic Upscaling of Heterogeneity

The translation of sorption data from the microscopic or laboratory level to the macroscopic or field scale level involves a number of assumptions and simplifications. Most of them are associated with the expression of local point estimates of flow and sorption into solute transport through field scale physical heterogeneity, notably defining statistical averages of geometry, physical and chemical characteristics, *etc.* The difficulties to predict sorption phenomena at the field scale level from rate coefficients derived at the laboratory level is one of the current main limitations in using non-equilibrium sorption models due to heterogeneities and structural patterns of the geological properties associated with successively larger scales.[41] By considering soils that have a narrow size distribution or soils that contain well-defined macropores, heterogeneity can be accounted for by dividing the porous media into mobile and immobile regions. The solute is transported by advection and dispersion in the mobile region and by diffusion in the immobile region. Its access to immobile regions is limited by diffusive mass transfer, which introduces the non-equilibrium transport behaviour. These two-region models have essentially the same mathematical structure as the two-site models and the analytical solutions are the same as long as the shape and size of the soil aggregates through which the solute diffuses is exactly known.

Mass transfer between regions can be described by diffusion equations,[1] on the assumption that the geometry of the immobile regions is known, or by first-order rate approximations.[42] Early diffusion-based models assumed that the soil particles were uniformly sized spheres but the applicability has been extended to non-spherical particles.[43,44] Moharir *et al.*[45] and Rasmuson[46] have developed models that include particle-size distributions, which tend to delay solute breakthrough in columns as a result of a high initial uptake rate of solute by smaller particles with relatively large surface areas. Wu and Gschwend[20] used a numerical method to examine the validity of using a mean size approximation to estimate sorption. A geometric mean size was a reasonable approximation when the size distribution was confined to one order of magnitude, but sorption was underestimated by a factor of two when the uniform size approximated an even size distribution spanning two orders of magnitude. The effect of particle size distribution on the mass transfer can normally be ignored if the coefficient of variation is less than 0.1[47] or if the range is less than 100 mesh.[48]

A mass transfer coefficient simplifies the expression of the solute transfer between the regions by relaxing the demands for a geometry description and by assuming that the transfer rate is a function of an average concentration difference between the mobile and the immobile regions. The rate coefficient is usually obtained by optimizing the fit of the transport model to the experimental data by a non-linear least-squares programme, but methods have been derived for independent measurements of the coefficient and for relating it to other measurable variables such as aggregate size and pore water velocity, which makes the mass-transfer model nearly as accurate as the diffusion-based model, at least for cylindrical macropore systems.[1,49] The complex mathematics involved in analytical solutions of the geometry-based models or the simplifying boundary conditions used in finite differences or finite elements numerical approximations have limited the applicability of the geometry-inspired models, and attempts have been made to define conditions for which simplifications are valid.[50]

6 Stochastic Approach

The scale dependency observed for hydraulic conductivity has been addressed by a stochastic approach, in which the hydraulic conductivity is represented in terms of random space functions, characterized by a limited number of statistical parameters.[51,52] As opposed to the deterministic approach, no detailed description of the geometry of the porous media is required. Similarly, the large spatial variability exhibited by sorption rate coefficients at the larger scales can be analysed by considering the interaction between the spatial distribution of sorption kinetics and the fluctuations of the flow velocity.[53] A common modelling approach is to assume a steady-state flow in an ensemble of homogeneous vertical stream tubes with flow properties varying randomly between the tubes[54,55] (Figure 4). It is generally agreed that the most important field scale variability of solute transport in both aquifers and unsaturated soils is due to spatial variation in hydraulic conductivity;[56–58] its coefficient of variation calculated from 17 field studies ranged from 0.46 to 6.27.[59]

With the flow velocities randomly varying, the expected concentration $\bar{C}$ at a given depth z can be expressed as

$$\bar{C} = \int_0^\infty C(z, t, D, v) P(v) \, dv \tag{8}$$

where $P(v)$ is the probability density function of the flow velocities. Under conditions of steady-state flow, the pore water velocity can be expressed as the saturated hydraulic conductivity, K_s, divided by the water content at saturation. K_s is usually regarded as the random variable with a log normal distribution. The macroscale transport process can be described by using either the resident concentration, that is, the mass of solute per volume of aqueous solution, or the solute flux concentration, that is, the average mass of solute per unit time crossing a unit area. Solute flux in a heterogeneous soil will result in a travel time distribution of solute reaching a certain depth in the soil at different times. The resident and flux average concentrations give similar estimates when the spatial heterogeneity is small but differ when the heterogeneity is large.[60] The stochastic travel time analysis can be combined with an expression for non-equilibrium

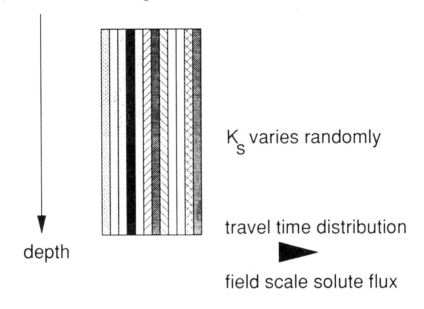

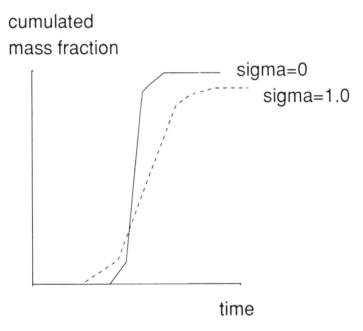

Figure 4 *In the stochastic approach to field-scale heterogeneity the vertical or horizontal flow field is represented by stream tubes, in which the hydraulic conductivity (K_s), sorption kinetics, etc. vary randomly, independently or correlated to each other. Heterogeneity can be observed from a travel time distribution translated into a solute flux description. The effect of variability (large sigma) in the physical parameter on solute transport can be demonstrated in cumulative breakthrough curves*

sorption kinetics assuming negligible local dispersion and molecular diffusion,[61] and the expected mass flux of the sorbing solute is defined by a probability density function of the travel time. This can be obtained either analytically by assuming some correlation structure between the sorption coefficients and the hydraulic conductivity,[62] or numerically by Monte Carlo simulations. When sorption kinetics are assumed to be first order with the forward and reverse rate coefficients equal and spatially constant, transport simulations indicate that both increased heterogeneity and sorption kinetics result in increased spreading and tailing of the expected solute flux.[63] When the sorption rate coefficients are spatially variable and either independent of or negatively correlated with the hydraulic conductivity, implying that low sorption rates are associated with high conductivity zones in the field, a double peak appears in the solute flux.[64] The dual peak behaviour of the stochastic travel time distribution has been demonstrated for, *e.g.* Br^- and Cl^- in well-structured field and laboratory soils.[65] It implies that solute flow is characterized by a fast component through interped voids and a slow component through intraped voids, but the stochastic analysis is unable to distinguish between solute retardation due to physical or chemical non-equilibrium phenomena.

7 Conclusions

The insufficient distinction between physical and chemical non-equilibrium is one of the problems limiting the understanding of solute transport in soil. Some resolution of that problem may be achieved by proper experimental manipulation of chemical and hydrodynamic conditions at the laboratory scale and identification of the correlation structures between local geochemical and hydrodynamic variables and a macroscale transport model. The spatial variation in sorption coefficients may reflect variations in the distribution and concentration of geopolymers, minerals, and charged constituents on the solids. A more detailed description of solute–sorbent interactions at the microscale level may help identify sorbent characteristics affecting sorption, especially in soils with low carbon concentrations and ionizable solutes. The main problem, however, is to re-analyse the current conceptual perspective on solute transport in heterogeneous soil and to design field experiments to test at least some of the embedded assumptions of the transport models.

References

1. P.S.C. Rao, D.E. Rolston, R.E. Jessup and J.M. Davidson, *Soil Sci. Soc. Am. J.*, 1980, **44**, 1139.
2. J.P. Gaudet, H. Jegat, G. Vachuad and P.J. Wierenga, *Soil Sci. Soc. Am. J.*, 1977, **41**, 665.
3. M.Th. van Genuchten, P.J. Wierenga and G.A. O'Connor, *Soil Sci. Soc. Am. J.*, 1977, **41**, 278.

4. M.Th. van Genuchten, 'Non-equilibrium Transport Parameters from Miscible Displacement Experiments', Research Report No. 119, USDA Salinity Laboratory, Riverside, CA, 1981.
5. W.C. Steen, D.F. Paris and G.L. Baughman, *Water Res.*, 1978, **12**, 655.
6. S.W. Karickhoff, D.S. Brown and T.A. Scott, *Water Res.*, 1979, **13**, 241.
7. W.P. Ball, C.H. Buehler, T.C. Harmon, D.M. Mackay and P.V. Roberts, *J. Contam. Hydrol.,* 1990, **5**, 253.
8. L.B. Barber II, E.M. Thurman and D.D. Runnells, *J. Contam. Hydrol.*, 1992, **9**, 35.
9. E. Tipping, *Geochim. Cosmochim. Acta*, 1981, **45**, 191.
10. J.A. Davis, *Geochim. Cosmochim. Acta,* 1982, **46**, 2381.
11. C.T. Chiou, L.J. Peters and V.H. Freed, *Science*, 1979, **206**, 831.
12. J.C. Means, S.G. Wood, J.J. Hassett and W.L. Banwart, *Environ. Sci. Technol.*, 1980, **12**, 1524.
13. R.D. Schwarzenbach and J. Westall, *Environ. Sci. Technol.*, 1981, **15**, 327.
14. E.M. Murphy, J.M. Zachara and S.C. Smith, *Environ. Sci. Technol.*, 1990, **24**, 1507.
15. D.R. Garbarini and L.W. Lion, *Environ. Sci. Technol.*, 1986, **20**, 1263.
16. T.D. Gauthier, W.R. Seitz and C.L. Grant, *Environ. Sci. Technol.*, 1987, **21**, 243.
17. P. Grathwohl, *Environ. Sci. Technol.*, 1990, **24**, 1687.
18. D.W. Rutherford, C.T. Chiou and D.E. Kile, *Environ. Sci. Technol.*, 1992, **26**, 336.
19. S.W. Karickhoff, in 'Contaminants and Sediments', ed. R.A. Baker, Ann Arbor Science Publishers, Ann Arbor, Michigan, 1980, Vol. 2.
20. S. Wu and P.M. Gschwend, *Environ. Sci. Technol.*, 1986, **20**, 717.
21. D.R. Cameron and A. Klute, *Water Resour. Res.*, 1977, **13**, 183.
22. P.S.C. Rao, J.M. Davidson, R.E. Jessup and H.M. Selim, *Soil Sci. Soc. Am. J.*, 1979, **43**, 22.
23. P.M. Jardine, J. C. Parker and L.W. Zelazny, *Soil Sci. Soc. Am. J.*, 1985, **49**, 867.
24. G.R. Southworth, K.W. Watson and J.L. Keller, *Environ. Toxicol. Chem.*, 1987, **6**, 251.
25. L.S. Lee, P.S.C. Rao, M.L. Brusseau and R.A. Ogwada, *Environ. Toxicol. Chem.*, 1988, **7**, 779.
26. L.S. Lee, P.S.C. Rao and M.L. Brusseau, *Environ. Sci. Technol.*, 1991, **25**, 722.
27. M.L. Brusseau and P.S.C. Rao, *CRC Crit. Rev. Environ. Cont.*, 1989, **19**, 33.
28. J.J.T.I. Boesten, L.J.T. van der Pas and J.H. Smelt, *Pest. Sci.*, 1989, **25**, 187.
29. J. Villermaux, *J. Chromatogr. Sci.*, 1974, **12**, 822.
30. H.M. Selim, *Environ. Health Persp.*, 1989, **83**, 69.
31. S. Akratanakul, L. Boersma and G.O. Klock, *Soil Sci.*, 1983, **135**, 267.
32. S. Akratanakul, L. Boersma and G.O. Klock, *Soil Sci.*, 1983, **135**, 331.
33. M.L. Brusseau, R.E. Jessup and P.S.C. Rao, *Water Resour. Res.*, 1989, **13**, 183.
34. P. Nkedi-Kizza, M.L. Brusseau, P.S.C. Rao and A.G. Hornsby, *Environ. Sci. Technol.*, 1989, **23**, 814.
35. A. Sabljić, *Environ. Sci. Technol.*, 1987, **21**, 358.
36. T.H. Carski and D.L. Sparks, *Soil Sci. Soc. Am. J.*, 1985, **49**, 1114.

37. D.M. Miller, W.P. Miller and M.E. Sumner, *Soil Sci. Soc. Am. J.*, 1989, **53**, 1407.
38. R. Salim and B.G. Cooksey, *Plant Soil*, 1980, **54**, 399.
39. P.M. Jardine and D.L. Sparks, *Soil Sci. Soc. Am. J.*, 1984, **48**, 39.
40. R.A. Ogwada and D.L. Sparks, *Soil Sci. Soc. Am. J.*, 1986, **50**, 1162.
41. C.J. Ptacek and R.W. Gillham, *J. Contam. Hydrol.*, 1992, **10**, 119.
42. M.T. van Genuchten and P.J. Wierenga, *Soil Sci. Soc. Am. J.*, 1976, **40**, 473.
43. P.S.C. Rao, R.E. Jessup and T.M. Addiscott, *Soil Sci.*, 1982, **133**, 342.
44. A. Rasmuson, *Chem. Eng. Sci.*, 1985, **40**, 1115.
45. A.S. Moharir, D. Kunzru and D.N. Saraf, *Chem. Eng. Sci.*, 1980, **35**, 1795.
46. A. Rasmuson, *Chem. Eng. Sci.*, 1985, **40**, 621.
47. D.M. Ruthven and K.F. Loughlin, *Chem. Eng. Sci.*, 1971, **26**, 577.
48. D.O. Cooney, B.A. Adesanya and A.L. Hines, *Chem. Eng. Sci.*, 1983, **38**, 1535.
49. I.S. Park, J.C. Parker and A.J. Valocchi, *Water Resour. Res.*, 1986, **22**, 339.
50. J.C. Parker and A.J. Valocchi, *Water Resour. Res.*, 1986, **22**, 399.
51. L.J. Gelhar and C.L. Axness, *Water Resour. Res.*, 1983, **19**, 161.
52. G. Dagan, 'Flow and Transport in Heterogeneous Formations', Springer-Verlag, New York, 1989.
53. V. Cvetkovic and A. Shapiro, *Water Resour. Res.*, 1990, **26**, 2057.
54. G. Dagan and E. Bresler, *Soil Sci. Soc. Am. J.*, 1979, **43**, 461.
55. W.A. Jury, *Water Resour. Res.*, 1982, **18**, 363.
56. L.W. Gelhar, *Water Resour. Res.*, 1986, **22**, 135S.
57. G. Sposito, W.A. Jury and V.K. Gupta, *Water Resour. Res.*, 1986, **22**, 77.
58. G.V. Wilson, J.M. Alfonsi and P.M. Jardine, *Soil Sci. Soc. Am. J.*, 1989, **53**, 679.
59. D. Isabel and J.-P. Villeneuve, *Ecol. Model.*, 1991, **59**, 1.
60. V.D. Cvetkovic and A.M. Shapiro, *Water Resour. Res.*, 1989, **25**, 1283.
61. J.-O. Selroos and V. Cvetkovic, *Water Resour. Res.*, 1992, **28**, 1271.
62. V.D. Cvetkovic and A.M. Shapiro, *Water Resour. Res.*, 1990, **26**, 2057.
63. J.-O. Selroos, Bulletin No. TRITA-VAT-1922, Royal Institute of Technology, Stockholm, Sweden, 1992.
64. G. Destouni and V. Cvetkovic, *Water Resour. Res.*, 1991, **27**, 1315.
65. R.E. White, J.S. Dyson, R.A. Haigh, W.A. Jury and G. Sposito, *Water Resour. Res.*, 1986, **22**, 248.

Natural Organic Substances and Contaminant Behaviour: Progress, Conflicts, and Uncertainty

By Angus J. Beck and Kevin C. Jones

INSTITUTE OF ENVIRONMENTAL AND BIOLOGICAL SCIENCES, LANCASTER UNIVERSITY, LANCASTER LA1 4YQ, UNITED KINGDOM

1 Introduction

Natural organic substances, of varying proportion and composition, are found in all soils and water (Chapters 1–3). They are generally more abundant in surface horizons of soils, where they are likely to interact with elemental, organic, or inorganic contaminants. These contaminants can enter soils through deliberate application, *e.g.* for the purposes of crop protection, inadvertently by chemical spillage, or naturally by atmospheric deposition. Interactions between naturally occurring organic substances and the contaminants of interest are of fundamental importance in governing contaminant persistence in soils and their potential to be transported to groundwater, and in some cases (*e.g.* non-ionic organic chemicals), those interactions may dominate their behaviour.[1]

In this chapter, we consider the 'state of the art', discuss continuing conflicts, and highlight remaining uncertainties in this field of research. In doing so we draw primarily on the earlier contributions to this book, although we also consider evidence from other sources. We do this as relative newcomers to the field, with a particular interest in the application of natural organic constituent–contaminant interactions to applied problems of land contamination.

2 'State of the Art'

The physico-chemical properties of organic matter in soils and its interaction with contaminants is now well understood at a general level. Traditionally contaminant behaviour has been predicted simply from knowledge of their distribution in a two phase system, *i.e.* solid and solution or soil and water. However, it is now widely recognized that if we are to reliably predict the behaviour of many contaminants in soils, it is necessary to have due regard to a third phase—dissolved organic matter (DOM).[2] Indeed, this may also be an oversimplification. We have seen in Chapter 2 that DOM (DOC—dissolved organic carbon, NOM—natural organic matter, WSSOM—water soluble soil organic matter) consists of a number of different subcomponents, *e.g.* hydrophilic and hydrophobic acid and neutral fractions, often defined on the basis of their extraction from water using the XAD series of macroreticular resins. Furthermore, it has now been demonstrated that these fractions are specific with regard to their interactions with both contaminants and soil constituents (organic and mineral) and thus their impact on contaminant behaviour is very complex (see Chapters 4–8). For many contaminants, particularly non-ionic organics, the presence of dissolved macromolecular substances has the effect of enhancing their solubility by stabilizing them in solution.[3–6] However, great care must be taken when resorting to such generalizations. For example, La France *et al.*[7] and Larsen *et al.*[8] have reported that even where interactions between dissolved organic matter and organic chemicals occurred, contaminant leaching was not significantly enhanced. The nature of solid organic substances has also been the subject of much research and many advances in our understanding have been made, such that Wershaw[9] has now postulated a model for humus structure. This model implies that humus comprises a combination of mineral grain coatings on the stationary phase and micelles or monomeric units in the solution phase. Both of these contain an internal hydrophobic region which is supple or 'liquid-like' and capable of conformational change, and an exterior surface which is highly charged.

Recent developments in spectroscopy and its increasingly widespread application to environmental problems has resulted in significant advances in our understanding of contaminant interactions with organic substances. Such studies have shown that a great diversity of possible interactions can occur for any contaminant in a given soil (see Chapter 4).[10] This in itself should not be surprising given the heterogeneous nature of both solid and dissolved organic substances in soils and water, and the number of reactive moieties on complex molecules such as pesticides. We have seen that the interactions occurring between contaminants and organic substances, be they stationary, mobile colloids, or soluble macromolecules, are essentially governed by the same physico-chemical principles. Consequently, it is possible to identify a relatively limited series of intrinsic characteristics of organic substances which can be used to predict their reactivity (Chapter 5). Kinetic constraints are now widely recognized as being of both great theoretical and practical importance. Pignatello (Chapter 6) has suggested that our interpretations of poorly understood phenomena such as the solids effect, isotherm non-linearity, and sorption to low organic matter solids may have been affected by our failure to completely understand the dynamic nature of the processes under investigation, and Mingelgrin and Gerstl (Chapter 5) have stressed that kinetic factors are partly responsible for our inability to predict sorption in a rigorous quantitative way.

However, it has now been demonstrated that stoichiometrically exact kinetic studies can be carried out in the laboratory that may yield parameters resulting in useful predictions of field scale behaviour (Chapter 7). From a practical standpoint, kinetics determine the long term persistence of contaminants in the environment and, consequently, they are likely to govern the extent to which contaminated land can be remediated in any given time period.[11]

Since the publication in 1989 of a frequently cited review by McCarthy and Zachara[2] on the sub-surface transport of dissolved organic matter and mobile colloids and their influence on contaminant transport, many more investigations have been carried out. Jardine and co-workers[12-14] have contributed much to our understanding of the mobilization and transport of organic matter in soils, and, in particular, its sensitivity to intrinsic environmental factors such as rainfall intensity and duration. The impact of dissolved organic matter and mobile colloids on the persistence and movement of a wide range of contaminants in soils including pesticides,[7,15-17] PAHs,[8] PCBs,[3,18,19] and chlorobenzenes[3,8] have also been investigated. Their role in groundwater transport is less well understood and requires further investigation.[20]

Not surprisingly, some modellers have taken advantage of these advances in their understanding to develop sub-routines for both existing and new models in the search for ever more reliable predictions of contaminant behaviour. Modelling contaminant interactions with solid phase organics has been discussed by Gamble and Khan[21] and dissolved organic matter by La France *et al.*[22] These principles have, in turn, been applied to field scale management and research models (*e.g.* Clemente[23]). The common approaches used were discussed in detail by Bengtsson in Chapter 9.

3 Continuing Conflicts

Modes of Contaminant Interactions with Organic Substances

One of the most emotive debates of the last decade has been the relative merits of partition and adsorption theory to explain the mode of interactions between non-ionic organic chemicals and organic substances. Essentially these two theories differ in that the former postulates interaction to be akin to the partitioning of a contaminant between a hydrophobic organic phase and an aqueous phase where only weak physical interactions occur. By contrast, adsorption involves subtle changes in the structure of the sorbent and sorbate when, for example, strong covalent bonds are formed between them. The evidence in favour of each theory has been presented and critically evaluated elsewhere[24,25] and is discussed briefly in Chapters 4 and 5. Evidence available to refute partition theory continues to mount, especially since spectroscopic techniques such as nuclear magnetic resonance and Fourier transform infra-red spectroscopy have become more widely used in the environmental sciences. Consequently, interest in the debate is now subsiding with the majority of the scientific community appearing to favour the assertion of Mingelgrin and Gertsl (1983)[25] that:

'Theoretically, or practically, surface uptake cannot be simply defined as 'adsorption' or 'partition', but rather there is a continuum of possible interactions starting with fixed site adsorption and ending with true partition between three-dimensional phases'.

Nevertheless, those researchers committed to partition theory are keeping the debate alive and regularly produce new evidence to substantiate their claims which provokes renewed interest as others seek to substantiate or refute this new evidence. Ultimately, this will only be beneficial in leading to a better understanding of the processes involved. A recent example has been the debate surrounding the surface area of organic matter.[26,27]

The Surface Area of Organic Matter

The surface area of organic matter is a critical concept for environmental chemists and engineers, because regardless of the mechanism of interaction between organic substances and contaminants, the absolute amount of surface area available will limit the sorption capacity of any chemical. It is therefore surprising that despite the enormous technological advances of recent decades, the 'specific surface area [*of organic matter*] remains an operational concept which is dependent upon the experimental method employed [*to measure it*]' Pennell and Rao, 1992.[26] These authors then go on to explain the effects of pretreatment such as oxidation and oven drying on measurements made using a variety of probe molecules including N_2, ethylene glycol (EG), *p*-xylene, toluene, and ethylene dibromide (EDB). For example, if the BET/N_2 method is used then estimates of surface area are generally much lower than if more polar molecules such as EG are used. On the basis of this evidence Pennell and Rao[26] proposed that organic matter comprises a very complex 3-dimensional matrix, with many reactive moieties and functional groups on both internal and external surfaces, such as those reported by Hayes *et al.*[28] and that proposed by Wershaw[9] which are susceptible to changes in their conformation and reactivity during hydration. By contrast, Chiou *et al.*[27] consider the model proposed by Pennell and Rao[26] to be overly complex and that the measures of 'effective' surface areas are exaggerated because the absorbates used as a probe (*e.g.* EG) frequently penetrate the solid itself, resulting in an overestimate of surface area based on 'inner surfaces'. Chiou *et al.*[27] have argued that the practical importance of this has not been adequately demonstrated. Thus, they have proposed that organic matter can be more correctly represented by a hydrophobic polymer phase on the basis of the small surface area of organic matter, determined by the BET/N_2 method.

Sorption–Desorption Kinetics: Elucidation and Evaluation of Rate Limiting Phenomena

Several mechanisms have been proposed to explain non-equilibrium sorption phenomena. They include transport-related non-equilibrium and sorption-related non-equilibrium.[29] Transport related (physical, flow-related) non-equilibrium occurs where the rate of sorption to soil solids and desorption from them is limited by pore water velocities in heterogeneous porous media.[30] These effects are well understood in a qualitative way, but less so in a quantitative sense. Nevertheless, their importance is widely recognized and little disputed. By contrast, sorption related non-equilibrium is less well understood and a number of different mechanisms are currently being postulated. These have been discussed in detail by Pignatello in Chapter 6 so only a brief overview and a summary of the major research needs are presented here.

Sorption-related non-equilibrium can be either chemical (chemisorption and physisorption) or physical (diffusive mass transfer). Chemisorption differs from physical non-equilibrium in that it is dependent on thermodynamic and kinetic controls and independent of physical processes such as diffusion of sorbates to sorption sites. Establishing the importance of chemical kinetics remains one of the most challenging problems in elucidating the mechanisms of slowly reversible sorption phenomena. The problem of doing so arises from the difficulty experienced in eliminating mass transfer constraints from experimental systems. Appropriate techniques have been developed by Sparks and co-workers[31,32] for extremely rapid reactions of metals on clay minerals and oxides, but they are highly specialized and have not yet been widely adopted or applied to other chemicals and soil components. Despite the partition debate, the importance of chemisorption with respect to the sorption of non-ionic organic chemicals requires further investigation because Senesi (in Chapter 4) has presented evidence which would suggest that covalent bonding between organic matter and many classes of organic chemicals is common. In Chapter 6 Pignatello argues that 'Although a contribution of adsorption to the kinetic component is conceivable . . . .' (citing a study by Szecody and Bales, 1989)[33] he concludes that 'it is difficult to believe that adsorptive forces alone could account for the long-term behaviour that is routinely observed'. Whilst this assertion may currently be appropriate, it may have to be revised on the basis of further spectroscopic and chemical kinetics research in the future.

Retarded intraparticle diffusion occurs when molecules are retarded due to reversible sorption on pore walls, whilst diffusing through water in micropores,[34] whilst intrasorbent diffusion occurs where resistance is encountered in three-dimensional polymeric organic matrices or expanding clay minerals. These two processes have frequently been confused and regarded as being synonymous in the literature. Brusseau *et al.*[29] have discussed the distinctions between these mechanisms and reviewed the evidence in support of each, concluding that intrasorbent diffusion (more specifically intraorganic matter diffusion) was primarily responsible for the sorption of a large number of hydrophobic organic chemicals including benzene, toluene, anthracene, naphthalene, chlorobenzene, tetrachloro-ethene, and quinoline in a fine sandy soil.

4 Remaining Uncertainties

Extraction, Analysis, and Structure of Naturally Occurring Organic Substances

Despite the tremendous progress that has been made in the development and refinement of analytical techniques over the last two decades and the consequential improvement in our understanding of the structure of naturally occurring organic substances, analytical chemists and surface scientists should not become complacent. They face many challenges in providing the information that is still required by those researchers interested in elucidating the mechanisms of interaction between contaminants and such substances and, in turn, those who are interested in the implications of such interactions at the field scale. The emphasis of investigations of the solid phase should focus on polysaccharides which are less well understood than the other components of organic matter. The proportion of humus which they constitute has yet to be reliably identified and their isolation from soils is required to facilitate in-depth study of their structural composition (see Chapter 3). A related objective should be the isolation and fractionation of plant root exudates (see Chapter 1) of which polysaccharides are believed to be a principal component. The difficulties of isolating such exudates in 'pure form' (*i.e.* as distinct from soil derived organic substances) represents the main obstacle to their investigation. However, these substances may be of importance with regard to contaminant uptake by plants. It is surprising, given the ease with which it can be sampled, that relatively little is known about the composition of dissolved organic carbon (DOC) in soil interstitial waters. Given that DOC may act as a carrier to facilitate the redistribution of some contaminants within soils, and that different fractions of DOC may interact with specific classes of contaminants its characterization must be rated as a high priority for future research (Chapters 1, 2, and 8). At present the extraction and fractionation of such DOC is a function of the methodology employed to date, *i.e.* chiefly retention on macroreticular resins. However, improvements in our understanding of contaminant transport and fate could be realized if research was geared towards the development of extraction methods which facilitate the identification of specific reactivities or behaviour (see Chapter 8).

Solubilization and Decomposition of Naturally Occurring Organic Substances

Long-term decomposition of organic matter on field plots has been extensively studied by Johnston and associates (see Chapter 1). They have found that despite receiving eight different organic ammendments (FYM, FYM compost, sludge, or sludge compost at 37.5 or $75\,t\,ha^{-1}\,yr^{-1}$) resulting in differences in the total humus contents of the experimental plots the rate of decomposition (1942–67) for all eight treatments was convergent on a single decay curve. This is an extremely interesting observation insofar as it suggests that the water soluble fraction of organic matter found in soil interstitial waters could be relatively uniform in its composition, regardless of the source of the organic matter from which it was derived (Chapter 1). This is perhaps at odds with the evidence presented by Malcolm in Chapter 2 which suggested that considerable variability is to be expected, and that uniformity of composition of DOC in soil interstitial waters and that extracted from humus, in the same locality, is the exception rather than the rule. Further research will be required if this paradigm is to be resolved and it should be apparent that naturally occurring DOC from soil interstitial waters rather than that extracted from solid phase soil organic matter should be used in any such research (Malcolm, Chapter 2). It seems likely that water soluble fractions of organic matter are more likely related to freshly added organic matter which is subject to relatively rapid microbial and chemical decomposition. This material may be derived from a whole range of different sources, including sewage sludge, crop residues, and farmyard manure, so a great diversity in composition might not be unexpected. By contrast, older more 'weathered' organic matter or humus is likely to be much more stable and less susceptible to solubilization relative to freshly added organic matter, and it may therefore be expected to have a relatively uniform composition.

Interactions Between Naturally Occurring Organic Substances and Contaminants

Although interactions between many classes of contaminants and naturally occurring organic substances are now well understood (see Section 2) challenging problems still remain. At a fundamental level, further spectro-scopic study can yield information that can enhance our understanding of the mechanisms involved and contribute to the resolution of those con-troversies discussed in Section 3. The need to develop appropriate methodological techniques which will facilitate the investigation of bound (stably-incoporated) residues has been identified by Senesi in Chapter 4. We also consider this to be of great importance because these residues currently limit the extent to which contaminated land can be remediated, and may represent an unspecified long-term risk to our environment. In both these regards, the rate at which contaminants are sorbed and desorbed is also of critical importance. For example, the success of bioremediation clean-up strategies for contaminated land is dependent upon the bioavailability of those chemicals giving cause for concern, which in turn may be rate limited by one or more of the kinetic constraints discussed in Section 3 (see Chapter 6). Thus, it is important that further research on these issues be carried out—not only in the interests of academic understanding, but to enable more effective management of our environment.

Mobility of Naturally Occurring Organic Substances and their Impact on Contaminant Transport

The potential for contaminants to be redistributed within soils and groundwater in association with dissolved organic matter or suspended particulate material is an extremely complex process. It is dependent on the multitude of possible interactions that can occur between naturally occurring organic substances and the mineral components of soils.[2] Although dissolved organic matter is known to be mobile, much research is still required to elucidate the mechanisms of solubilization (as detailed previously) and, once in solution, the reimmobilization as coatings on clay mineral surfaces (Chapter 8). In Chapter 5 Mingelgrin and Gerstl have demonstrated that a few simple physico-chemical rules effectively govern all the interactions between organic substances and contaminants, but have also stressed the limitations of using such a unified approach. The same principles are likely to apply to the interactions that occur between dissolved organic substances and stationary organic matter or clay minerals. McCarthy (Chapter 8) has shown that different fractions of dissolved organic matter move at different rates through soils, and that these fractions may be specific with regard to the classes of contaminants with which they can interact. Much basic research is still required to determine the effects of the shape and reactivity of the molecules, and steric factors on the mobility of dissolved organic matter in soils. At the larger scale, further carefully controlled field experiments are required to ascertain the overall impact of mobile organic matter on contaminant transport in both saturated and unsaturated porous media. However, given the complexity of the soil system, we appear to be destined to largely rely upon indirect inferences from simpler laboratory experiments.

Modelling the Behaviour of Naturally Occurring Organic Matter and Contaminants in Soils

Although it is now widely recognized amongst environmental chemists and engineers that dissolved organic matter is of great importance with regard to the environmental behaviour of many classes of contaminants, parameters representing this impact have not yet been routinely incorporated in mainstream management or research transport and fate models. Thus, one challenge for the environmental chemists is to persuade the model builders of the need to do so, if only for the purpose of demonstrating whether or not such additional complexity is manifest in more robust prediction. However, in so doing it will be necessary for chemists to appreciate that 'specifics of interaction' may need to be sacrificed in favour of 'generalities of behaviour' if models are to find general application. This is not to say that there is no place for more detailed mechanistic models such as the stoichiometrically exact approaches of Gamble *et al.* (see Chapter 7). Such models are extremely useful insofar as they contribute greatly to our understanding of parameter sensitivity at the local scale, and if they can be demonstrated to provide reliable predictions, their incorporation into larger scale models for the management of major site-specific environmental problems may prove to be very beneficial.

5 Conclusions

It should be apparent from the foregoing discussion and preceeding chapters that if we are to reliably predict the behaviour of contaminants in soils and groundwater, the role of natural organic substances cannot be ignored. Although much has been achieved in the extraction, isolation, and fractionation of organic substances in soils and water which has led to the elucidation of their molecular structures, researchers should not become complacent. Much can still be gained from such fundamental research if the unresolved problems identified in Section 4, *e.g.* polysaccharides and hydrophobic fractions, are given research priority. The application of spectroscopic techniques, *e.g.* NMR, to the study of contaminant inter- actions with both solid and dissolved macromolecular organic substances show much promise. Whilst this is only likely to fuel rather than resolve the continuing adsorption *versus* partition debate, the ongoing interest can ultimately only be beneficial to our understanding of the processes taking place. The need to adopt kinetic approaches to study such interactions in preference to the current preoccupation with equilibrium-based studies is now apparent. This is particularly true where the behaviour of interest is over the longer term at the field scale, where mass transfer constraints, at both the microscopic and macroscopic scales, may be rate limiting.

Regardless of the scale of interest, the need to adopt multidisciplinary or unified approaches to the problem of concern is now widely recognized. These approaches should be encouraged, because they are most likely to lead to significant advances in our understanding. For example, at the local scale little can be gained from conceptualizing a system in two domains, *i.e.* soil and water, where three or more domains are appropriate. At the larger scale, *e.g.* soil monoliths or field studies, the same reasoning applies. However, to be of practical use we must ultimately be able to express our understanding of the processes taking place in a quantitative way. In doing so it will be necessary to restrict the number of model parameters to an absolute minimum, and it is also desirable that such parameters should be both directly and easily measured. It is this requirement to reconcile complex physico-chemical processes in both space and time with simple mathematical expressions that poses the greatest challenge to contemporary research. Ultimately, it may be necessary to sacrifice 'specifics of inter-action' for 'generalities of behaviour', which will provide a good approxi-mation of contaminant behaviour under typical conditions, but may prove to be inadequate if predicting unusual events giving rise to the pollution incidents which often give us most cause for concern.

Acknowledgements

We are grateful to the UK Ministry of Agriculture, Fisheries and Food, the Natural Environment Research Council and the Agricultural and Food Research Council for supporting our research on the fate and behaviour of contaminants in soils.

References

1. A.J. Beck, A.E. Johnston and K.C. Jones, 'Movement of Nonionic Organic Chemicals Through Agricultural Soils', *Crit. Rev. Environ. Sci. Technol.*, 1993, **23**.
2. J.F. McCarthy and J.M. Zachara, *Environ. Sci. Technol.* 1989, **23**, 496–502.
3. C.T. Chiou, R.L. Malcolm, T.I. Brinton and D.E. Kile, *Environ Sci. Technol.*, 1986, **20**, 502–508.
4. G. Caron, I.H. Suffet and T. Belton, *Chemosphere*, 1985, **14**, 993–1000.
5. S.J. Traina, D.A. Spontak and T.J. Logan, *J. Environ. Qual.*, 1989, **18**, 221–227.
6. DY. Lee and W.J. Farmer, *J. Environ. Qual.*, 1989, **18**, 468–474.
7. P. La France, L.Ait-ssi, O. Banton, P.G.C. Campbell and J-P. Villeneuve, *Water. Pollut. Res. J. Can.*, 1988, **23**, 253–269.
8. T. Larsen, T.H.Christensen, F.M. Pfeffer and C.G. Enfield, *J.Contam. Hydrol.*, 1992, **9**, 307–324.
9. R.L. Wershaw, *Environ. Sci. Technol.*, 1993, **27**, 814–816.
10. N. Senesi and Y. Chen, 'Interaction of Toxic Organic Chemicals with Humic Substances', in 'Toxic Organic Chemicals in Porous Media', Ecological Studies 73, eds. Z. Gerstl, Y. Chen, U. Mingelgrin and B. Yaron, Springer Verlag, Berlin, 1989, pp. 37–90.
11. US Environmental Protection Agency, Summary Report High-Priority Research on Bioremediation, Bioremediation Research Needs Workshop, Washington DC, 1991, p. 7 cited in J.J. Pignatello, Chapter 6, this volume.
12. P.M. Jardine, N.L. Weber and J.F. McCarthy, *Soil Sci. Soc. Am. J.*, 1989, **53**, 1378–1385.
13. P.M. Jardine, G.V. Wilson and R.J. Luxmoore, *Geoderma*, 1990, **46**, 103–118.
14. G.V. Wilson, P.M. Jardine, R.J. Luxmoore and J.R. Jones, *Geoderma*, 1990, **46**, 119–138.
15. J.C. Madhun, J.L. Young and V.H. Freed, *J. Environ. Qual.*, 1986, **14**, 64–68.
16. D-Y. Lee, W.J. Farmer and Y. Aochi, *J. Environ Qual.*, 1990, **19**, 567–573.

17. E. Barriuso, U. Baer and R. Calvert, *J. Environ. Qual.*, 1990, **21**, 359–367.
18. J.P. Hassett and E. Millicic, *Environ. Sci. Technol.*, 1985, **19**, 638–643.
19. F.M. Dunnivant, P.M. Jardine, D.L. Taylor and J.F. McCarthy, *Environ. Sci. Technol.*, 1992, **26**, 360–368.
20. D.A. Backhus and P.M. Gschwend, *Environ. Sci. Technol.*, 1990, **24**, 1214–1223.
21. D.S. Gamble and S.U. Khan, *J. Agric. Food Chem.*, 1990, **38**, 297–308.
22. P. La France, O. Banton, P.G.C. Campbell and J-P. Villeneuve, *Sci. Total Environ.*, 1989, **86**, 207–221.
23. R.S. Clemente, 'A Mathematical Model for Simulating Pesticide Fate and Dynamics in the Environment (PESTFADE)', PhD Thesis, MacDonald Campus, McGill University, Montreal, QC, Canada, 1992.
24. C.T. Chiou, L.J. Peters and V.H. Freed, *Science*, 1979, **206**, 831–832.
25. U. Mingelgrin and Z.J. Gerstl, *J. Environ. Qual.*, 1983, **12**, 1–11.
26. K.D. Pennell and P.S.C. Rao, *Environ. Sci. Technol.*, 1992, **26**, 402–404.
27. C.T. Chiou, J-F. Lee and S.A. Boyd, *Environ. Sci. Technol.*, 1992, **26**, 402–404.
28. M.H.B. Hayes, P. McCarthy, R.L. Malcolm and R.S. Swift, 'Humic Substances II. In Search of Structure', John Wiley and Sons, New York, 1989.
29. M.L. Brusseau, R.E. Jessup and R.S.C. Rao, *Environ. Sci. Technol.*, 1991, **25**, 134–142.
30. M.L. Brusseau, *J. Contam. Hydrol.*, 1992, **9**, 353–368.
31. D.L. Sparks, *Advan. Agron.*, 1987, **38**, 266.
32. D.L. Sparks, 'Kinetics of Soil Chemical Processes', Academic Press, London, 1989.
33. J.E. Szecsody and R.C. Bales, *J. Contam. Hydrol.*, 1989, **4**, 181 cited in J.J. Pignatello, Chapter 6, this volume.
34. S. Wu and P.M. Gschwend, *Environ. Sci. Technol.*, 1986, **20**, 717–725.

Subject Index

Acetone, 96, 133
Acetonitrile, 133
Actinides, 161, 168
Activated carbon, 77
Activation energy, 135
Acylation, 80
Adsorption bond energy (ABE), 133–135
Adsorption isotherms, 82, 103–107
Adsorption mechanisms, 81–90
Adsorptive capacity, 154, 157
Advection-dispersion-reaction (ADR), 129, 131, 161, 174
Aerial deposition, 12, 73
Alachlor, 80, 84, 89
Aldicarb, 79, 80
Aldrich humic acid, 135
Aldrin, 79
Algal blooms, 14
Alkanes, 137
Alkylbenzenes, 137
Alkyl-phenyl silica, 135, 136
Alkylation, 80
Alkylnaphthalenes, 56
Alkylphenanthrenes, 56
Aluminium, 167
Americium, 168
Amines, 34, 87
Amino acids, 48, 49
Amitrole, 77, 84, 85
Anion exchange, 155
Anion-exchange chromatography, 42
Anionic compounds, 78
Anthracene, 51, 189
Apparent sorption coefficient (K_{app}), 130, 162
Apparent rate coefficient, 177
Aromaticity, 22, 23, 28, 157, 165, 173
Asymmetric breakthrough curves, 171, 174, 176, 177
Asulam, 84

Atrazine, 91, 92, 95, 109, 115, 116, 130, 133, 145, 146, 149

Background descriptive variables, 144
Basic compounds, 77, 78
Batch sorption techniques, 155, 158, 163, 173, 177
Bentonite, 115
Benzene, 114, 189
Benzene polycarboxylic acids, 52
Benzidene, 87
Benzo[a]pyrene (BaP), 132, 165, 166
Benzoate, 35
Benzonitrile, 89
Benzophenone, 96
BET isotherms, 104
BET/N_2, 187, 188
Bioaccumulation, 74, 76, 77
Bioavailability, 74, 102
Biodegradation, 74, 80, 132
Bioremediation, 133
Blue dextran, 156
Boron trifluoride–methanol, transesterification, 55
Bound residues, 190
Bovine serum albumin, 162
Bromacil, 84
Bromide, 130, 156, 181
Bromoxynil, 78
Butralin, 80

Cadmium, 12, 13, 18, 163–165, 176
Calcium chloride, 15
Carbaryl, 79, 89
Carbofuran, 79, 80
Carbon dioxide, 3, 27, 49
Carbon:nitrogen ratios (C:N ratio), 7, 17, 74, 75
Carbonic acid, 3
Carbothioate herbicides, 80

Carboxyl groups, 23, 26, 33, 43, 47, 48, 52, 53, 76, 78, 82, 86, 91, 143, 145, 146, 165
Cation exchange, 82–84
Cation exchange resin, 20
Cationic compounds, 77, 82
Cations, 142
CDAA, 80
CDEL, 80
Charcoal, 11
Charge transfer, 84–86
Chemical kinetics (chemical non-equilibrium), 143, 158, 171, 181, 188
Chitin, 57, 130
Chloramben, 78
Chloranyl, 86
Chlordane, 79
Chlordimeform, 82, 86
Chloride, 27, 28, 130, 134, 159, 181
Chlorinated benzenes, 129, 186, 189
Chlorpropham, 79
Coal, 11, 23
Colloids, 80, 102, 117, 153–155, 167, 168, 185, 186
Column capacity factor (K'), 21
Copper, 88, 115, 144
Covalent bonding, 86–88, 186, 188
Crop residues, 190
Cycloate, 80, 84, 89
Cycluron, 79

Dalapon, 78
DBCP (1,2-dibromo-3-chloropropane), 130
DDT, 79, 86, 89, 90–92, 94, 102
Dechlorohydroxylation, 91
Decomposition, 7, 17, 23, 190
D_{eff} (effective diffusion coefficient), 131, 136, 137
Detergents, 77
Detoxification, 74
Diazinon, 79, 91
Diaminoethane (DEDA), 36
Diborane, 43
Dicamba, 78, 84, 89
Dichlobenil, 80
Dieldrin, 79
Diethylaminoethyl (DEAE) cellulose, 42

Differential thermal analysis (DTA), 84
Diketones, 94
Dimefox, 84
N,N-Dimethylformamide (DMF), 35
Dimethylsulfoxide (DMSO), 35, 38, 65
Dinoseb, 78
Dioxins, 86, 93
Diphenamide, 80
Diphenols, 84
Dipole–dipole interactions, 89, 135
Diquat, 77, 82, 86
Dissolved organic carbon (DOC), 19–21, 185
instrumental analysers, 19, 20
Dissolved organic matter (DOM), 130, 135, 185
Disulfoton, 94
Diuron, 79, 110, 111
Dry combustion, 11

Electron donor–acceptor interactions, 84–86
Electron microscopy, 32
Electron spin resonance, spectroscopy (ESR), 85–88
Electrophoresis 32, 41, 58
Electrostatic factor (EF) values, 34
Elemental carbon, 11
Empirical models, 141
Enzymatic catalysis, 86, 87
Enzyme-reaction kinetics, 171
Endrin, 79
Energetics, 108–110
Enolate anions, 34
Enols, 33, 39
Entropy, 55
Ethanoate, 35, 43
Ethene, 189
Ethylene dibromide (EDB), 118, 132–134, 157
Ethylene glycol (EG), 187
Exchangeable cations, 111, 112
Extraction,
of humic substances, 33–37
of polysaccharides, 37–38

Facilitated transport, 162
Fonofos, 82

Farmyard manure, 5, 6, 8, 9, 11, 13–15,190
Feldspar, 173
Fenuron, 79, 80, 86
Fertility, 3, 65
Fertilizers, 5, 6, 9, 10, 11, 14, 73, 141
First-order kinetics, 92, 95, 129, 130, 136, 174, 178, 181
Fluorometuron, 80
Fluorene, 111, 118–120
Fluoridone, 119
Fly ash, 33
Formic acid, 35
Fourier transform infrared spectroscopy, 32, 167, 186
Fractionation,
    of humic substances, 38–41
    of polysaccharides, 41, 42
    of soil organic matter, 33–43
Free radicals, 36, 76, 85, 87, 88, 95
Freundlich equation, 82
Frictional ratio, 60
Fucose, 57
Functionality,
    of humic substances, 43–48
    of polysaccharides, 48
    of soil organic matter, 42–48
Furan, 55
Furfural, 55

Galactosamine, 57
Galactose, 57
Galacturonic acid, 57
Gas chromatography, 32, 53, 55
Gel permeation chromatography, 39–42, 58
Geopolymers, 173
Glucosamine, 57
Glucose, 56
Glucuronic acid, 57
Glycosidic linkages, 48, 56, 61
Glyphosate, 79, 84
Groundwater, 20, 23, 24, 28, 32, 73, 74, 132, 153, 155, 158, 163–165, 167, 168

Halogenated hydrocarbons, 130, 136
Heats of adsorption, 155
Heptachlor, 79
Herbage, 12, 13

Hexane, 108, 111, 115, 119
Histosols, 24
Hyamine, 82
Hydrochloric acid, 37
Hydrogen, 24, 34, 43
Hydrogen bonds, 34, 35, 65, 84, 135, 144
Hydrogen peroxide, 93, 134, 177
Hydrolysis, 48, 49, 73, 77, 80, 91, 92
Hydrophilic acids, 20, 21, 24, 26
Hydrophilic bases, 21
Hydrophilic neutrals, 21, 22, 24, 28, 29
Hydrophobic acids, 20, 21, 157, 165
Hydrophobic bases, 20, 21
Hydrophobic neutrals, 20, 21, 22, 24, 28, 29, 157, 165
Hydrophobic partitioning, 138
4-Hydroxybenzenecarboxylate, 35
Hydroxy radicals, 93
Hydroxybenzene, 44, 51
Hysteresis, 117, 122

Inceptisol, 24
Incineration, 73
Industrial waste, 73
Inner variables, 143
Interstitial water, 23–29, 189, 190
Intraorganic matter diffusion, 133, 134, 136, 137
Intrapolymer diffusion, 174, 177
Intrasorbent diffusion, 171, 188
Ion exchange, 175
Ionic boundary, 82–84
Iron, 88, 167, 168
Irving–Williams stability constants, 145
Isomorphous substitution, 124
Isopropanol, 115

Labile sorption, 141, 148
Landfill effluent, 73
Langmuir equation, 82, 104–107
Leaching, 6, 10, 11, 15, 17, 27, 134, 185
Lead, 13
Leptophos, 89
Ligand exchange, 88, 155
Lignin, 50, 54
Lignite, 23
Lindane, 78, 79, 90, 130

Linear free-energy relationship
  (LFER), 135, 136
Linuron, 79, 82
Lipids, 89
Lithium, 34
Local equilibrium assumption (LEA),
  128, 171, 177

Macroreticular resins, 185, 189
Macroscopic scale, 172, 178, 192
Macrospecies [definition], 102
Magnesium, 3
Magnetic species, 108
Malathion, 79, 84
Mannose, 57
Marine DOC, 27, 28
Mass-action expression, 103–107, 143
Metals, 12, 13, 32, 141, 144, 145, 161,
  165, 167, 188
Methazole, 89
Methanol, 133
Methiocarb, 79
Methoxyl groups, 23, 86
1-Methylnaphthalene, 135
Methyl parathion, 86
Methylbenzene, 56
Methyl cyanide, 34
Methylene blue, 96
Methylfuran, 55
Methylfurfural, 55
Metolachlor, 84, 130, 133, 135
Micelles, 90, 185
Microbial activity, 8, 18
Microbial biomass, 8
Microbial degradation, 171, 190
Microscopic scale, 172, 177, 178, 192
Mineralization, 11, 14, 17, 32
Miscible displacement, 177
Modelling, 191–193
  deterministic approach, 178, 179
  sorption, 173–176
  stochastic approach, 171, 179–181
  two-phase approach, 129, 130,
    136, 173, 174, 176, 177
Moieties, 35, 84, 89, 110, 185
Moisture, 8, 116–121
Molecular connectivity indices, 109
Molecular size and shape,
  humic substances, 58–61
  polysaccharides, 61

Mollisol, 24
Monte Carlo simulation, 181
Montmorillonite, 112, 114
Monuron, 79
Mössbauer spectroscopy, 44
Mulches, 4
Multilayer sorption, 104

Naphthalene, 189
α-Naphthylamine, 87
Napropamide, 105–107, 115, 116, 130
Neptunium, 168
Net adsorption energy concept, 108
Nitralin, 80
Nitrate, 10, 14, 17
Nitrogen, 3, 5, 6, 10, 11, 14, 24, 26,
  32, 36, 76, 84, 187
Natural organic matter (NOM), 153
Non-equilibrium sorption, 128, 155,
  157, 171–173, 176
Non-ionic compounds, 78–81
Non-linear least-squares, 178
Norflurazon, 89
Nuclear magnetic resonance,
  cross-polarization magic angle
  spinning (CPMAS), 22, 24, 26–28,
  32, 44–46, 75, 186, 192
Nucleic acids, 48, 49

Orizalyn, 80
Osmometry, 58
Outer variables, 143
Oxalate, 35
Oxidative degradation processes,
  51–53
Oxides, 33, 142, 154, 155, 167, 168,
  188
Oxygen, 24, 26, 76, 84

Paraquat, 77, 82, 86, 94, 102
Parathion, 79, 86, 89, 108, 115
Particulate carbon, 19
Partition coefficient, 121–123, 129,
  137, 138, 162, 177
Partitioning, 73, 89, 90, 121–123, 163,
  186, 187, 188, 192
Pentachlorophenol (PCP), 78
Pentafluorobenzoate ion, 137
Peptide bonds, 48, 49
Peroxy radicals, 93, 94

Persulfate, 27
Pesticides, 19, 73, 74, 77, 86, 87, 141–144, 185, 186
Petrol, 77
pH, 12, 13, 21, 36, 39, 74, 76, 82, 90, 92, 111, 115–116, 138, 143, 144, 145, 167
Phenacridane chloride, 77, 84
Phenate anions, 34
Phenol degradation processes, 53–55
Phenolic groups, 35, 43, 64, 76, 78, 82, 86, 88, 92, 173
Phenols, 33, 47, 78, 87, 39
Phenolic peaks, 23, 24
Phenoxyalkanoic acids, 78, 82, 88, 89, 91
Phosphon, 77, 82, 84
Phosphorus, 3, 13–15, 18, 32, 174
Photodecomposition, 73, 93–96
Photoejection, 93
Photolysis, 86
Photosensitization, 93–96
Phthalic acid diesters (PAE), 77, 81, 90
Physical heterogeneity, 171–173
Physical non-equilibrium, 188
Phytotoxicity, 73, 79
Picloram, 78, 82, 88, 89
Plastics, 77
Plutonium, 168
Polychlorinated biphenyls (PCBs), 86, 89, 90, 102, 163–165, 186
Polyclar - AT, 37
Polyenes, 94
Polymer diffusion, 121
Polymeric colloids, 138
Polynuclear aromatic hydrocarbons (PAHs), 77, 80, 81, 90, 94, 129, 132, 165, 186
Polysaccharide components, 56, 57
Pore diffusion, 133, 134
Pore water velocity, 159, 172, 176
Potassium, 3, 34, 35
Potassium hydroxide, 36
Preferential adsorption, 108, 155, 157, 164, 168
Preferential flow, 155, 158
Probability density function, 181
Profluralin, 80
Prometryne, 84

Propachlor, 80
Propazine, 91
Propham, 79
Pseudopolysaccharides, 55
Pump-and-treat remediation, 132
Pyrazone, 80
Pyrolysis, 55, 56
Pyrolysis mass spectroscopy (PyMS), 32

Quartz, 173
Quinoline, 189

Radial diffusion, 136
Radionuclides, 161, 165
Radius of gyration, 61
Raman spectroscopy, 44
Random space functions, 179
Reductive degradation processes, 51
Resistant sorption, 128, 129
Retarded intraparticle diffusion, 141, 146, 148, 149, 188
Rhamnose, 57
Rhizosphere, 8
Riboflavin, 96
Ribose, 57
Root exudates, 8, 189
Rose bengal, 96
Rothmund–Kornfeld equation, 82
Run-off, 73

Saturation limits, 103, 111, 143, 164, 165, 187
Sephadex gels, 40–42
Sewage sludge, 8, 15, 190
Silica gel, 135
Simazine, 11, 91, 130, 132
Singlet oxygen, 93, 94
Size-exclusion effects, 156, 161
Smectites, 114
Sodium, 34
Sodium hydroxide, 36, 37
Sodium salts, 35, 36
Sodium sulfate degradation, 53–55
Soil texture, 7, 9, 17
Solubility enhancement, 90, 91, 153, 185
Solubility parameter theory, 108
Solubilization, 73, 90, 91
Solvated electrons, 93

Solvents, 77
Solvophobic interactions, 108
Sonication extraction, 133
Sorption capacity (see saturation
  limits)
Sorption–desorption kinetics,
  analytical methodology, 133
  bioavailability, 132
  contaminant transport, 129–131
  longterm persistence, 132
  mechanisms, 133–137
  remediation, 132, 133, 190, 191
  research needs, 137, 138
Sorption-related non-equilibrium, 188
Sorption-retarded pore diffusion, 134,
  137
Soxhlet extraction, 133
Spatial variability, 158, 171, 172, 178,
  179, 179
Speciation, 76, 144, 161
Steric effects (see also size exclusion
  effects), 56, 112–116, 119, 124, 128,
  136, 191
Stoichiometry, 141, 142, 186
Streams, 21–23, 27, 29, 73
Substituted benzenes, 135
Sub-surface transport (see leaching)
Succinate, 35
Sugar acids, 26
Sulfuric acid, 37
Sulfur, 76
Superoxide radicals, 93, 94
Superphosphate, 12, 14, 15
Surfactants, 77, 132

2,2,5,5-TCB, 166
3,3,4,4,-TCB, 166
TeCB (1,2,4,5-tetrachlorobenzene),
  130
Temperature, 8, 32, 90, 138
Temporal factors, 112–116
Tetrachloroethane (PCE), 130, 135,
  189

Thermodynamics, 90, 129, 134, 138
Thiocarbonate, 80
Thyamine, 77
Titration, 47, 48
Toluidine, 87
Toluene, 130, 187, 114
Total organic carbon (TOC), 19
Toxaphene, 79
Transport-related non-equilibrium, 188
Triazines, 77, 78, 82, 84–86, 88, 89,
  91, 136, 137
Trichloroethene (TCE), 130, 134
Trifluralin, 80
Trihalomethanes, 19
Triton X-100, 162
Trona waters, 23

Ultracentrifugation, 32, 58
Ultrafiltration, 39, 41
Ultra violet–visible spectroscopy, 44
Uranium, 168

Vadose zone, 156
Van der Waals interactions, 65, 89,
  135
Viscometry, 58
Volatilization, 76, 79, 80

Weighted averages, 143, 165
Wet oxidation, 27

X-ray diffraction, 33
X-ray photoelectron spectroscopy, 44
XAD-4 acids, 21, 22, 24, 26
XAD-8 fraction, 161, 165
XAD-4 resin, 21–23, 28, 38, 39
XAD-8 resin, 20, 21, 28, 37–39, 49,
  55, 157
Xylene, 130, 187
Xylose, 57

Zero point of charge (ZPC), 173
Zonal centrifugation, 39, 41